DEFINING SCIENCE

RHETORIC OF THE HUMAN SCIENCES

General Editors

John Lyne
Deirdre N. McCloskey
John S. Nelson

Defining Science

A RHETORIC OF DEMARCATION

CHARLES ALAN TAYLOR

THE UNIVERSITY OF WISCONSIN PRESS

The University of Wisconsin Press
114 North Murray Street
Madison, Wisconsin 53715

3 Henrietta Street
London WC2E 8LU, England

5 4 3 2 1

Printed in the United States of America

Library of Congress Cataloging-in-Publication Data

Taylor, Charles Alan.
Defining science: a rhetoric of demarcation /
Charles Alan Taylor.
302 pp. cm. — (Rhetoric of the human sciences)
Includes bibliographical references and index.
ISBN 0-299-15030-5 (alk. paper) .— ISBN 0-299-15034-8 (pbk.: alk. paper)
1. Science—Philosophy—Case studies. I. Title. II. Series.
Q175.T37 1996
501'.4—dc20 96-180

CONTENTS

ACKNOWLEDGMENTS

As with all phenomena, a book such as this is the product of profound interconnectedness, of relationships between me and countless others. Among those I can count, however, the following deserve special mention. First, my teachers. From my undergraduate education, I owe a debt to Darrell Langley, Jim Zabel, and Bill Trollinger, whose insightful and supportive guidance suggested to me the value of the "life of the mind." At the University of Illinois, Celeste Condit, Dan O'Keefe, Barb O'Keefe, Jesse Delia, David Swanson, Cheris Kramarae, and Andy Pickering gave me invaluable insights and support. Special thanks and respect are due to Joe Wenzel, my doctoral adviser. His rigorous, compassionate, and unwavering support illustrated graphically what a scholar and friend can be.

Jim Andrews, Pat Andrews, Mike Hogan, Bill Wiethoff, Cherie Bayer, Nicki Evans, and Carolyn Calloway-Thomas, my colleagues at Indiana University, deserve thanks for making a young scholar feel at home and for crafting an environment in which this work came to fruition. I especially value the guidance and support of Bob Ivie and John Lucaites, who supported this project when I wasn't sure it had a future and supported me when I'm not sure I deserved it. I also want to thank the many students with whom I worked at IU, especially Jim Cherney, who served admirably as a research assistant. Our department's crack administrative staff, Helen Harrell, Bonnie Clendening, and Laura Gano, have my undying gratitude for all their help with this project.

Many others have contributed to this work through their suggestions, criticisms, and general "smartness." I especially want to thank Steve Fuller, John Nelson, John Lyne, Larry Prelli, Alan Gross, and Leah Ceccarelli for their insights. Rosalie Robertson and the editorial staff of the University of Wisconsin Press have been models of goodwill and efficiency, characteristics all too uncommon among university presses.

On a personal note, my parents, Jim and Mary Taylor, deserve my respect and gratitude for their support and much more. Finally, this book is dedicated to Jan and Megan: for everything and nothing in particular.

DEFINING SCIENCE

1 THE RHETORICAL ECOLOGY OF SCIENCE

First, a few words about definition in general. Let's admit it: I see in a definition the critic's equivalent of a lyric, or of an aria in opera. . . . In actual development, the definition may be the last thing a writer hits upon. Or it may be formulated somewhere along the line.
—Kenneth Burke, *A Grammar of Motives*

What then is science? If no one asks me, I know; but if I want to explain it to a questioner, I do not know.
—Max Charlesworth, *Science, Non-science and Pseudo-science*

In a 1938 letter to the Italian minister of state, Albert Einstein protested plans to require Italy's intelligentsia to pledge loyalty to the fascist regime. He wrote that "the pursuit of scientific truth, detached from the practical interests of everyday life, ought to be treated as sacred by every Government, and it is in the highest interests of all that honest servants of truth should be left in peace" (1990, 32). In his honorific prose, Einstein articulated a view of science as a practice whose uniqueness from other pursuits, most notably politics, demanded special, even deferential treatment. Science was not, he made clear, a political enterprise. Left unsaid, of course, was what science *was*. While his rhetorical task presumably did not require his taking a position on that issue, it is nonetheless striking that he urged an exemption from the oath only for a social group whose membership was based on traits whose ontological and evaluative solidity was taken wholly (and fawningly) for granted.

In his 1953 Reith Lectures for the BBC, J. Robert Oppenheimer crafted the metaphor of science as a house whose sheer magnificence rendered trivial any particular blueprint. As he put it, "It is a house so vast that there is not and need not be complete concurrence on where its chambers stop and those of the neighboring mansions begin" (1966, 84). One might wonder, though, how the owners of those neighboring abodes would react to Oppenheimer's definitional and architectural egalitarianism.

These two anecdotes point toward the intellectual and practical difficulties of identifying those characteristics of science that define it as a discrete social practice, distinguishing it from neighboring mansions, competing intellectual practices, or pretenders to its epistemological throne. Sir Karl Popper described this task, of providing a "definition of the idea of an empirical science" (1959, 42), as "the *key* to the most fundamental problems of the philosophy of science" (1962, 42).

It seems clear that the *demarcation* problem, motivated by our urge to define *science,* has been seen simultaneously as of considerable importance and as quite nearly intractable. Indeed, the intellectual horizon is littered with attempts to come to grips with the constitutive character of science. With few exceptions, the goal of myriad demarcation discourses tends toward the explanatory (Holton 1988), a construct invariably embedded into larger epideictic portrayals of an intellectual practice as the penultimate expression of the Enlightenment. This celebratory bent is made all the more striking by the fractious, inconsistent, and fundamentally inconclusive ontological judgments made regarding the constitutive nature of the object of celebration. Perhaps, as Burke suggests, definitions of science and humankind simply get "formulated somewhere along the line."

Given the seeming inability of any particular demarcation tale to account adequately for the nature of science and the purported irrelevance of that account for the doing of science (Holton 1988), we might be tempted to turn for answers to those who ought to know best—scientists themselves. After all, science might well be understood as "what scientists do." Pickering, arguing generally in this spirit, construed science as "a way of being in, getting on with, making sense of, and finding out about the world" (1992, 19–20). Who would be better trained to provide answers to the demarcation question than those who get on with, make sense of, and find out about the world? In short, the scientists themselves are the most reliable sources on the nature of their activities. While myriad commentators report on science post facto, the nature of the activity itself can be taken for granted, as a sort of tacit knowledge, or at best, we are left to take scientists' accounts of their practices as isomorphic with those practices, effacing, as it were, the map and its territory.

The View from Here: Demarcation as Rhetorical Practice

At the risk of plunging headfirst into tautology, I want provisionally to endorse the dictum "Science is as science does"—though this endorsement does not extend to the nascent ontologies enacted in extant discussions of demarcation. In this study, I will tell a quite different story about science, one in which the taken for granted (the elusive, but more or less stable nature of science) is problematized, and the previously problematized (its cultural significance) is taken generally for granted. Unreflective commitments regarding the heretofore "unproblematic" issue lead, I want to argue, to unfortunate enactments of the problematic. This inversion of problematics is my initial step toward a rhetorical perspective on the "demarcation" of science. From this perspective, the discursive practices of multiple social actors, including but hardly limited to practicing scientists, are taken as constructing the boundaries that mark off the domain of science from, for example, pseudoscience and politics. In such symbolic processes of marking off, discursive practices craft and sustain formative links with practices typically construed (by traditional approaches to the demarcation problem) as lying outside the boundaries of science. The multiple entailments of those interconnections are absolutely central to the technical practices of contemporary science, as well as to our ability to integrate those practices into the fractured deliberations characteristic of the postmodern polis.

The view of science developed here holds that the meaning of science, as a set of social practices, is constructed in and through the discourses of scientists as they respond rhetorically to situations in which certain of their social, technical, professional, and technical interests are problematized in ways that may or may not be apparent either to the scientist or to the analyst of her or his activities. Practicing scientists, consciously or otherwise, discursively construct working definitions of science that function, for example, to exclude various non- or pseudosciences so as to sustain their (perhaps well-earned) position of epistemic authority and to maintain a variety of professional resources. Departing significantly from previous approaches to demarcation that presume a single boundary between science and "nonscience," "metaphysics," or "religion," this perspective illustrates how competing research communities within traditional science construct interdisciplinary demarcations so as to advance proprietary interests over particular research domains and/or experimental practices and exemplars. As a consequence, these symbolic practices can be read as defining implicitly what it means to do science as myriad social actors go about the professional activities of funding, research, review and publication, error correction, and utilization of the products of contemporary science.

Perhaps the central insight of a rhetorical analysis of demarcation practices appears, at first glance, rather transparent: science and, by implication, scientific knowledge are social constructions which are given presence in rhetorical discourse. Science is what it is rhetorically demarcated and authorized to be. A commitment to a view of scientific practice as defined symbolically is unequivocally, even shamelessly, fragmented. It is a view that celebrates the heterogeneity of science as it is played out through the interconnections of practices that constitute it. As Jasanoff has observed in a different context, multiple social actors "have a stake in the definition of science . . . and the vocabulary used by all of these parties remains . . . 'essentially contested'" (1987, 224).

As a conceptual and methodological assumption, this is hardly an endorsement of Pyrrhic relativism for its own sake, for the traditional "debunking" imperative need not obtain at this level. Rhetorical critics, concerned generally with the discourses practically tied to communal life, would seem to be in an unusually fruitful position to exploit both the undeniably useful productions of science as well as the inexorable elasticity of the multiple discursive formations constituting science and the broader formations with which they interpenetrate. This is not to say that science is somehow a product of blind luck (be it good or bad). Clearly, historically productive patterns, norms, and assumptions do constrain the future discursive outlines of the culture as they accumulate epistemic and practical presumption. This presumption, however, is itself elastic. It does not dictate the shape or rule out any potential mutation of contemporary practices. The stabilization of these discourses into what canonical histories of science portray as the "essence" of science is a powerful rhetorical accomplishment. For example, the degree to which any particular configuration opposes the domains of science and politics, via a language that "exports politics out of science" (Latour 1990, 147), should not be taken as a necessary polarity, but rather as a consequence of symbolic inducement.

The particular contours of a contemporary culture of science are not products of, as Rorty (1987) puts it, a "natural kind." Nor should we assume them to be "unnatural kinds." Indeed, it seems preferable to take science as heterogeneous kinds that stabilize in particular configurations as a consequence of the symbolic mediation and transformation of multiple constraints. From such a view, emphasizing the interdependent cooperation (not always cooperative, mind you; indeed, usually contentious) of multiple discursive formations, science can be seen as a motile, universalizable category, but one that, to borrow Leff's description of rhetoric, "finds its habitation only in the particular" (1987a, 7).

Accordingly, we must abandon the simplistic belief in science as somehow nonconstructed, as a set of practices or knowledge claims whose defining essence is constituted uniquely beyond the realm of practical human activity. Rorty's admonition is suggestive: "We need to stop thinking of science as the place where the human mind confronts the world and of the scientist as exhibiting proper humility in the face of superhuman forces." What is needed, as Rorty phrases it, is a "way of explaining why scientists are . . . exemplars which does not depend on a distinction between objective fact and something softer, squishier, and more dubious" (1987, 39). Presumably, the softer, squishier, and more dubious is thought to be the realm of the social, the cultural, the rhetorical. Abandoning such reductive opposition, as this view of demarcation does, allows a more meaningful understanding of the interpenetration of the symbolic and material. Science, then, is neither a mirror of nature nor intellectual anarchy. Rather, it is a production of historically shifting social communities that rhetorically demarcate themselves from other communities.

On a practical level, this orientation broadens the range of social practices that are potentially interconnected within the discursive formations of science and, hence, that function as sites for demarcation practices. Many such practices might heretofore have been considered nonscientific, or, at best, peripherally so. Science here consists not simply of laboratory practices and their finished products, but also of the wide-ranging activities that facilitate those and other practices, including securing funds, ensuring public patronage, and the like. "Quark-barrel" politics, then, is as much a fact of scientific life as it is a fact of political life on Capitol Hill. Such a commitment does not require abandoning a concern with how science "actually gets done." On the contrary, it broadens the range of formations in which we can observe, evaluate, and perhaps even enhance that discursive "doing."

The notion of a scientific *ecosystem* might be illustrative on this point. Containing numerous constituents, an ecosystem recognizes the primacy of certain species within their ecological niches. That primacy, however, comes not as a function of one species' isolation from others, but from the ecosystem's profound interconnectedness.[1] Just as Ameri-

1. As I will illustrate in subsequent pages, this metaphor enacts perhaps the most significant distinction between my account of demarcation and the extant literature, namely that the identity of science is a product not of its ontological difference, but rather of its contextual connectedness. What science is (speaking nonontologically), then, is a conse-

can upper-class taste for ivory accessories was tied inextricably to the near extinction of entire species of elephants in the nineteenth century, so decisions of congressional committees to make possible (or impossible) particular research ventures are tied to eventual judgments regarding the potential facticity of the phenomena under investigation.

Alternately, the constituents of the natural and scientific ecosystems are not identical in authority, nor do they operate from identical goal orientations. The redwoods of the Pacific Northwest do not share motives with Weyerhauser, nor is the level of power maintained by government (de)regulators shared by environmental activists. In like fashion, the interests of fusion physicists need not be shared by entrepreneurs interested in commercial applications of cold fusion technology, and the power granted to funding agencies is not shared by hopeful researchers or fiscal watchdog groups. Hence, particular hierarchies are the products of complex natural processes in the natural world and of multiple rhetorical processes in the cultural ecosystem.

The myriad constituents of our ecosystem exist in a fluctuating equilibrium. Rhetorically tipping the balance in one direction or another will send reverberations throughout the system. Systemic imbalances in favor of northwestern timber interests or of Earth First! would alter irrevocably the fragile ecosystem, to the potential detriment of the larger whole.

How the practices of any single species affect myriad others is the outcome of contextual interactions, emphasizing the profound interconnectedness of the whole. For example, the recent identification of the so-called breast cancer gene might be read as one particularly fortuitous interaction of research scientists, federal granting agencies, pharmaceutical industry concerns, and women's health advocates, constrained as well by individual rivalries and institutional alliances within biomedical research communities. Similarly, consider the configuration of sociotechnical practices that led to the demise of the Superconducting Supercollider. The supercollider began its public life as something of a scientific panacea, enabling the pursuit and potential capture of the elusive "top quark," hence reinforcing the prestige of "American" science. As the project ran into inevitable cost overruns, delays, and partisan bickering, however, this monument to the scientific mind underwent a metamorphosis into a "shameless boondoggle," designed merely

qence of particular, historical episodes in which the constituents of the ecosystem stabilize. Naturalist John Muir, for example, described one motivation for his environmental activism: "When we try to pick out anything by itself, we find it hitched to everything else in the universe." Of course, such a perspective appears counterintuitive on its face and, admittedly, raises a host of intellectual problems of its own.

to reward influential members of Congress or the pampered elites of high-energy physics.

The constituents of the scientific ecosystem (understood as the complex, but open-ended, network of cultural practices) function similarly. Tipping the balance solely in favor of "science in its own terms" points toward a corrosive variety of "scientism" (Sorrell 1994). On the other hand, opting for government microintervention in technical decisions conjures frightening recollections of Nazi racial science.

While the particular means to this end is the truly difficult question, the more or less peaceful and balanced coexistence of multiple species is the goal toward which we ought to strive. Just as Oregon's experience since 1992 illustrates that jobs and the spotted owl can coexist, profitably in both cases, my account calls for a recognition of the inescapable symbiosis of what traditionally we have called the technical and the public, the internal and the external, or the natural and the cultural.

This is not simply an intellectual exercise, at least not in the typical, banal sense of that phrase. Indeed, as we increasingly confront the discursive manifestations of the postmodern age, it is difficult, if not impossible, to tell with confidence the difference between the expert and the crackpot. The key, as I see it, is not to seek refuge in too-easy fundamentalisms, be they religious, scientific, or deconstructionist, but to exploit the generative capacity of rhetorical encounters to produce "good reasons" on which our collective judgment might be based.

Why a Rhetorical Perspective on Demarcation?

As I suggested earlier, most demarcation discourses reflect the disciplinary position and interests not so much of practicing scientists, but of philosophers. A staple of philosophical controversy for centuries (Sorrell 1994, 2ff.) and perhaps cast in its most common contemporary form by Popper, the demarcation question has been answered in a host of inconsistent ways. Unifying those otherwise quite disparate formulations, however, has been a general commitment to the citation of various epistemic markers that delimit the nature of science as a means to the "preservation of scientific truth" (Amsterdamski 1975, 127). Larry Laudan (1983), for instance, has argued that a philosophically significant demarcation criterion must distinguish scientific and nonscientific enterprises in a way that "exhibits a surer epistemic warrant or logical ground for science than for non-science" (18). Hence, the sine qua non of most philosophical accounts has been the ultimate convergence of ontological and normative claims regarding science.

Historians and sociologists of science (and/or scientific knowledge) appear fundamentally concerned to identify recurrent social structures or sets of knowledge claims that are characterized as scientific within particular historical contexts and to account for the social influence constraining those characterizations. Only occasionally assuming the epistemic superiority of science, sociological accounts have been cast as features of the conduct of the scientific community (Merton 1973) and/or its constituent members (Gieryn 1983a).

Rhetorical scholars have only recently (and tentatively) begun to speculate on the rhetorical dynamics of demarcation (Holmquest 1990; Lessl 1988; Taylor 1991b). This study is an attempt to fill that critical void, not by suggesting a rhetorical quick fix to a vexing intellectual problem, but by bringing the theoretical and methodological skills of rhetorical criticism to bear on the demarcation narratives told both by scientists and by those who make a living listening to them, observing them, or even following them around, as Latour (1987) might put it.

Let's return momentarily to the examples mentioned earlier. A rhetorical analysis of the alleged isolation of the "breast cancer gene" would be well positioned to map the discursive means by which the multiple discursive formations in which technical practices were enacted intersected, and to what end. Similarly, it could illuminate the arguments around which the individual and institutional alliances were created and upon which the "ecosystem" achieved its local, temporary, and hardly inevitable "closure" in announcements of the gene's identification. In the case of the supercollider, a rhetorical analysis would not be concerned with the ontological status of the "top quark," but with the rhetorical stabilization of this fractious scientific ecosystem and its implications for understanding the shifting nature and authority of contemporary big science—issues that might themselves constrain ontological judgments.

The trajectory of my text is built around a series of embedded narratives, drawn from the multiple disciplines that have dealt with a rhetorical account of demarcation. In chapter 2, I take up the tortuous paths followed by philosophy in its search to find the "grail" of demarcation, one that protects science from its contenders or, worse yet, its pretenders. Those paths veer from the avowedly prescriptive to less formalistic description, and they bring us to the edge of social epistemology. Chapter 3 picks up the story at the sociological border and recounts the journeys launched into the urban jungles of science in search of their constitutive nature. From social structures to social constructions to social discourse, we are brought face to face with the domain of the symbolic, the rhetorical. Chapter 4 engages the tales told regarding science by scholars of rhetoric. Hacking our collective way through epistemic thickets, not to mention various rhetorical contexts, communities, and spheres, we find ourselves in a position to reflect backward, freely re-visioning narratives so as to warrant a rhetorical perspective on demarcation. That perspective then allows us to make sense of "real" scientific demarcation controversies.

Using our rhetorical perspective as a compass, chapter 5 offers a detailed reading of the contemporary controversy over creationism, calling attention to the multiple demarcations accomplished with varying degrees of authority between professional science and its public constituencies. Chapter 6 interrogates the implicit demarcation rhetorics at play in the recent furor over the alleged discovery of "cold fusion." Serving not to secure the perimeter of science against incursions from the public, this tale reveals the reification of a communal identity, *via negativa,* in the process of communal error correction.

Given the relative novelty of my interpretive commitments, a few comments regarding my reading strategy might be in order here. Initially, my analyses of the philosophical and sociological literatures do not suggest that scholars from Popper to Fuller, from Merton to Latour, have *really* been doing rhetoric all along but simply haven't recognized it. Put another way, I do not suggest that there is a nascent "rhetoricism" in their work that can simply and unproblematically be translated into some sort of rhetorical account.[2] Mine is a quest for reconstruction, not redescription. As a consequence, I read this literature through rhetorical lenses. Here, then, I will be reading consistently against the grain, frequently against the sources' "preferred interpretations," and perhaps a bit uncharitably. Nonetheless, I want to see how far this interpretive churlishness might take us as we map the discourses of demarcation—a strategy that prefigures any future reconstructions.

It should be clear that I do not attempt to answer the questions posed in these narratives on their terms, or even to answer these questions at all. Rather, my strategy is to glean from these interdisciplinary accounts of demarcation insights which can be coupled with distinct lines of work in rhetorical studies so as to craft a unique and useful perspective

2. Gaonkar (1990, 1993) has chided a number of "rhetoric of science" scholars for the seeming arrogance of assuming an implicit and unrecognized rhetorical turn in "master texts," from Kuhn's *Structure of Scientific Revolutions* to Habermas' *Legitimation Crisis.* While simply mapping a rhetorical vocabulary onto insights generated elsewhere is parasitic and derivative, importation and exportation of intellectual capital offer a valuable opportunity to craft more inclusive accounts of difficult issues.

for understanding demarcation. It remains, however, a *perspective,* a subjective position from which to view a similar set of social practices. There are, of course, other perspectives, and the views from those perspectives might be quite different.

This raises two important issues. The first is that a rhetorical perspective on demarcation is not concerned with the ontological solidity of the boundaries drawn discursively between science and its symbolic neighbors. I make no claim that the "real" nature of science is located properly within only a limited range of interconnected practices. One errs, at least in the subjective position of a rhetorical critic, in assuming that the telos of rhetorical studies of science is to elucidate the irreducible character of science, to go where legions of other scholars have gone before only to fail. Beyond revealing appalling disciplinary hubris, such a stance presumes that *the* answer is out there like Higgs's elusive Boson waiting to emerge from behind unworthy discursive formations. Such an a priori assumption, it seems to me, is a uniquely academic case of the tail wagging the dog.

Just as interpretive heterogeneity marks my retelling of the "metadiscourses" about science, so too it structures my engagement of the texts of science per se. Perhaps more than in any other intellectual practice in history, scholars have tended to tread lightly where the technical proclamations of science are concerned. After all, we compose at advanced personal computers and travel to conferences on jets based on the work of physicists and other scientists, not rhetorical critics. While I do not challenge the utility of such work, I fear that letting that overdetermine our interpretive strategies has produced what Fuller aptly labeled "a passive consumer culture that has no formal mechanism to account for what scientists do" (1993, 381). This "psychosis," I think, is a function of the pernicious assumption that the veracity of a given disciplinary critique of a discourse practice is to be evaluated fundamentally by its relative ability to be recognized by the discourse practitioner, that is, that it reflected what he or she "really meant." Surely, to the extent that we claim (misleadingly, I think) to offer accurate descriptions (or transcriptions), we assume the burden of accuracy.

Extending that standard to other critical practices, however, utterly eviscerates the power of critique and destroys its reformative potential. Perhaps we ought not disavow the "unintended," or even the "counterintuitive," as worthy characteristics of our critical readings. Perhaps it is within such "alienated" accounts that, as Gergen put it, "should the occasional player enter into such a reality, his/her posture toward the activity might never be the same" (1992, 366). To the extent that our readings are "absorbed" into the putative accounts of our objects of analysis and critique, they (and we) forfeit any meaningful opportunity for remaking the cultural formations in which our professional and public lives are embedded.

This is an especially poignant issue for rhetorical criticism, interested as it has traditionally been in the discourses of the polis. Those discourses, concerned in their broadest senses with matters of how we ought to formulate the discursive formations within which our professional and public selves are enmeshed, demand from us a reconstructive move. It is one thing to recognize explicitly, as I do, that the texts of rhetorical critics (or any other professional texts) are themselves self-interested enactments of ideological constraints and, thus, to call for their reflexivity (Ashmore 1989).[3] It is quite another, however, to allow that call to paralyze the critic's right or responsibility to argue for the position he or she advances. For the rhetorical critic, language is more pragmatic than deconstructionism's caricatured "free play of the signifier," far more complex and interesting than an "open sesame" to rhetors' cognitive processes, and far more, in almost every way, than just talk (Myers 1989). It is, perhaps more than anything else, the currency of our cultural transactions. As a consequence, the "critical rhetoric" I seek to develop here self-consciously *and* reflexively suggests alternative symbolic constructions to those it deconstructs.[4]

3. The issue(s) of reflexivity—in ethnographies, in critiques, in linguistic forms of all sorts—has emerged as an important concern of the humanities in general and science studies in particular (e.g., Gruenberg 1978; Babcock 1980; Ashmore 1989; Pollner 1991). The tenor of this debate is perhaps best captured in Woolgar's wry observation that "relativism brings out the religious in people. Reflexivity, it seems, brings out the venom" (1988a, 430). To make an unbearably long story mercifully short (itself a tactical textual move), the controversy involves the relative authority of a given textual rendering and renderer. If we insist on the indeterminate textuality of our objects of study, how can we make meaningful our critiques, insofar as they too must exhibit this indeterminate status? Several quite creative textual maneuvers, for example, explicitly arguing against the claims made in a given text (Ashmore 1989), have sought to enact reflexivity. While useful, such playful moves, it seems to me, might be rendered more so if one concedes indeterminacy, advancing an argumentative position, all the while holding that position open as precisely that—a product of argument. As with all arguments, then, it remains subject to reconsideration and rejection. The sophistic roots of the rhetorical tradition, for example, would seem to point in just such a direction, empowering both deconstruction and subsequent reconstruction.

4. By invoking the notion of a critical rhetoric, I am identifying with recent speculations within rhetorical studies. Drawing from Foucault's analysis of power/knowledge, McKerrow's articulation of critical rhetoric advances simultaneous critiques of power domination and freedom. The critique of domination seeks to "undermine and expose the discourse of power in order to thwart its effects in a social relation" (McKerrow 1989, 98). The corresponding critique of freedom enables the critic to advance alternative symbolic construals, predicated on the freedom *to* (e.g., take action) and freedom *from* (e.g., from op-

I am concerned here to speak from a vantage point that must become more pronounced as the constituents of the scientific ecosystem spread ever more widely throughout contemporary culture—that of a member of the body politic, neither Luddite nor techno-geek. As a consequence, I do not claim possession of technical training in, say, fusion physics or paleoanthropology. My interpretations are, of course, open to challenge on multiple grounds, including "empirical" ones. However, the fact that they are not products of advanced scientific training is, it seems to me, a necessity, even a virtue, rather than a limitation or a vice. To the extent that the increasing ambiguity and importance of negotiating the boundaries of what we take to be science now place a larger burden on the "public sphere," taking the scientific community, in Polanyiesque fashion, solely on its own terms seems an inappropriate interpretive (and political) strategy. As the new millennium approaches, any meaningful "knowledge policy" (Fuller 1993) will require meaningful public participation in decision domains too long allocated solely to science per se. The profoundly embedded and tangled sociopolitical webs in which science gets done are graphic illustrations of the need for interpretive approaches to science that do not start by privileging only those texts that scientists (or traditional philosophers or historians) would label appropriately scientific (Dickson 1988, 306ff.). To suggest otherwise, it seems to me, would be to dull the critical edge so crucial to recent developments in "science studies," and even in "science" itself.

The periodic labeling of rhetorical studies as interloper or dilettante, occasioned by the perception that it dallies with other discourses with only minimal concern for their empirical integrity, is not a phenomenon limited to philosophers and/or scientists. Within rhetorical studies, in fact, Leff (1987b) and McGuire and Melia (1989) have suggested a tension between rhetoric's traditional concern with the polis and the avowedly insular character of technical rhetorics. While I question the necessity of adhering to traditional disciplinary frameworks as historical epochs evolve, there seems a more direct and compelling reason to avoid suc-

pression). An interesting series of critical engagements has isolated a number of lacunae in McKerrow's original articulation (e.g., Hariman 1989; Ono and Sloop 1992), cohering generally around the related issues of agency and telos. In short, these critics suggest that incessant critique leaves the reconstructive "dream" incessantly "deferred," with the postmodern indeterminacy foreclosing the ground for advocacy. This, it seems to me, assumes, mistakenly, that advocacy requires foundational grounds. On the contrary, it is the recognition of discursive indeterminacy that opens the space for both critique and reconstruction, so long as the terms of that reconstruction are argued for, rather than assumed. While typically not cast in such terms, the project of critical rhetoric itself is an enactment of a practical (albeit temporary) resolution of the "problem" of reflexivity—conceding its necessity, and then "getting on with it."

cumbing to such fears. Most notably, the practical demarcation between the activities of "big science" and the polity blurs increasingly in the (post)modern age. From questions regarding whether and where to locate toxic waste dumps to the requirement of citizen jurors to make sense of often contradictory "expert scientific" testimony, the fruitful exercise of *phronesis* draws increasingly (and critically) on our ability to make meaningful judgments of technical rhetorics. As a consequence, the reformative potential of a rhetorical perspective on science requires that the opaque discursive formations of science become "equipment for living," not only for those most centrally located in them, but also for those whose lives will, for better or worse, be transformed by them. It is that audience of which I am a member, and to whom I communicate.

The perspective I will develop is concerned fundamentally with the functional use of discourse to define, redefine, even to deconstruct, the implicit boundaries of those social practices we consider scientific. This is not to say that science is reducible to rhetoric, or that science is necessarily rhetorical, all the way down (Gross 1991; McGuire and Melia 1991). Rather, it is to suggest that the contextually variable configurations of practices that constitute science are defined rhetorically for its constituent members, and significantly, the points of interface with other social practices are identified rhetorically. As a result, the various components of the scientific "ecosystem" are both defined and animated by "symbol (mis)using creatures," in Burke's ([1968] 1994) terms.

Therefore, we *can* tell where science ends and society begins, and vice versa. We do it every day. Few of us, for example, expect, less yet desire, the next Nobel Prize in physics to be awarded to an avant-garde performance artist, or even to a physicist cum sociologist of scientific knowledge. Such demarcations of science are accomplished routinely in everyday social and scientific practice. What must be recognized, though, is that such demarcations proceed not from ontological foundations but from symbolic inducements. They are, then, rhetorical accomplishments. In this way, a rhetorical perspective on demarcation illustrates the practical epistemic utility of contemporary science as well as the constructive authority of rhetorical discourse. Of course, there are other practices with epistemic utility and other practices with constructive authority. They, though, are another story.

What this perspective forces us to take seriously is the prospect of a truly constructive rhetoric of science (Peters and Rothenbuler 1989). Recent developments in the so-called rhetoric of inquiry (e.g., Lyne 1985; Nelson, Megill, and McCloskey 1987; H. Simons 1985) have begun to indicate that discourse is a constitutive factor in the conduct of scholarship and public affairs. Indeed, it is not going too far afield to argue that the rhetoric of inquiry movement is itself an exercise in demarcation, to the extent that it views discourse as the organizing construct around which academic disciplines can be allied and distinguished (e.g., Leff 1987b).

There is certainly some merit to the argument that science, as a group of disciplines or intellectual endeavors, addresses certain topics and makes certain methodological choices that other intellectual endeavors tend not to make (or at least do not make as often). It does not seem entirely correct, however, to reify those choices as concretely defining science. To do so would belie the heterogeneity of human practices in general and science in particular. Such a reification indirectly legitimates a conception of science as somehow privileged above those intellectual endeavors which are only more obviously constrained by social and rhetorical factors. If we are led unreflectively to sanction quasi-objectivist demarcations of science, we necessarily distance ourselves from the material conditions of their production and legitimation.

Of particular concern would be those situations in which the constituent nature and practices of science are called into question. By unreflectively positioning particular "norms" or "criteria" as constitutive of science without regard for the particularities of their rhetorical articulation and deployment in specific controversies, we risk erecting a monolithic epistemic front where one does not, cannot, and should not exist.

This emphasis on scientific controversy entails a commitment to empirically relevant rhetorical theory. The rhetorical view of demarcation outlined here and applied in the upcoming case studies is grounded in historically material examples of demarcation. Demarcation, then, is not a phenomenon (rhetorical or otherwise) which is somehow divorced from the conduct of scientific activity. Indeed, it is a central (constitutive) constraint on the everyday productive and valuative practices of scientists as laborers. The rhetorical process of demarcation is part of the "work of science." The principles and norms which have traditionally been said to define science are meaningful only in social action, in praxis. To suggest otherwise, I think, is to maintain the mystification which has for too long empowered uncritical claims to epistemic authority.

In short, then, this perspective calls for a recognition of the inescapable symbiosis of what we have traditionally kept separate by artificially rigid demarcation criteria. To the extent that we understand science as having nontrivial interfaces with public and political practices, we can observe and comment critically on the particular configurations in which these interfaces are rhetorically structured.

This shift in perspective affords us the opportunity of regrounding scientific practice into a broader social analysis in which the constraints of ethical and political practice reflexively influence the nature and operation of scientific practices. Barnes and Edge have argued that "the knowledge and culture of science are developed . . . in specific contexts and how far those people keep . . . distinct from and unconditioned by the broader social context is a wholly contingent matter, needing separate ascertainment in each case" (1982, 188). While ascertainment might be a difficult critical standard to meet, the cultural cartography accomplished in demarcation practices is a central constraint on just how far the culture of science, inextricably bound to the cultural world, attempts to articulate its "distinctness" from that world. Contrary to the traditional approach to demarcation, in which science is operationalized via its relative conceptual difference, its identity (understood rhetorically) is a localized consequence of its connectedness to other social practices.

Accordingly, this perspective moves beyond the simple oppositions of scientific versus nonscientific or technical versus public to encompass a broader constructive understanding of all knowledge production. This enables us to interrogate the rhetorical construction of science as well as its emergent relationships with the broader contexts in which it is embedded and over which it is too often accorded epistemic privilege.

While not everyone will be comfortable with the potential open-endedess of the tale outlined here, it should be emphasized that this position need not lead us into a vicious relativism (McGuire and Melia 1991) or a myopic antiscientism (Myers 1990). Indeed, it need not raise epistemological questions at all. Peters and Rothenbuler suggest that the rhetorical approach offers a path beyond objectivism and simpleminded relativism. They argue, "Let us do away with both the positivist's horror . . . and the debunker's glee and come to terms with what we as a species have *made* and are *making*. The symbolic performative and discursive production of reality is . . . a given of human society and experience" (1989, 24).

The task, then, is to move beyond this claim to a consideration of the way in which we define as scientific certain of those processes in which realities are rhetorically produced and sustained. This is a complex question, requiring simultaneous regard for the ontological commitments of our main characters, as well as for the primary influences of the rhetorical constraints on the enactment of those commitments. Lyne (1990a) has wisely counseled that "rhetoric alone . . . cannot substitute for all the methods, plodding, tools, rationales, and observables that the various sciences depend upon" (55). In short, rhetoric is not everything that scientists do. Nonetheless, this story calls attention to the cultural rhetorics upon which those very "observables" depend.

At this juncture, the sine qua non of contemporary culture, its politics, is located at the heart of all discursive practices, including demarcation discourses. Insofar as scientific ways of knowing are generally granted epistemic authority in contemporary culture, interrogating the situationally emergent standards by which they are effectively demarcated from other ways of knowing offers potential for revealing the complex relationships between knowledge and the structures of social power (Aronowitz 1988a, 1988b; Schiappa 1989).

The insistence that science and its knowledge products are social constructions leads inexorably to a commitment to the centrality of human factors in the hierarchical construction and legitimation of knowledge-producing enterprises. David Miller argues that we should "regard ideas as tools with which social groups may seek to achieve their purposes in particular situations. . . . The conflict between interests and ideas is contextually mediated" (1986, 228). The point at which human factors are incorporated into a social conception of knowledge production is the point at which particular patterns of knowledge may become reconceptualized as, to some extent, a social resource. Bohme and Stehr have argued that "knowledge is a social characteristic . . . i.e., something which patterns the social behavior of its carriers, constitutes peculiar social relations, serves as a basis of social status and a resource for living" (1986, 53). Knowledge, particularly of the sort associated with science, takes on the role of a commodity, cultural goods to be produced, accumulated, and consumed. Consequently, cognitive barriers erected according to the relative degree of legitimacy accorded prevailing scientific paradigms are effectively translated into "socio-cultural barriers" which empower the "true believers" and subjugate the "uninformed" (Dolby 1982, 267).

This is not to say that our goal must be to eliminate such usages as a means to the end of reclaiming an appropriately *objective* science. At the very least, such a goal reflects a psychotic nostalgia for a time that never existed. The interpenetration of the cultural and scientific is hardly something we must guard against; by recognizing the contingency of such a barrier, we can work toward the edification of both. The practical successes of science are undeniable and profound. As Barber has indicated, "Through all the multifarious benefits it has brought, science has increased public trust in its competence . . .. and in its fiduciary responsibility for the public welfare" (1987, 132). After all, few of us place faith in the physics of rhetorical criticism as we board an aircraft.

Nonetheless, it would be a mistake to underestimate the hegemonic potential of particular constructions of science, even if those constructions are founded upon declarations of nonconstructedness. Extending his critique to the political economy of science, Aronowitz observed that "the scientific and technological revolutions today constitute the hegemonic culture of advanced industrial society. The discourse of conservatism and reaction are dependent on science . . . to persuade an entire population . . . or to impose themselves as necessary discourses" (1988a, 535).

It is not my intention here to launch another salvo in this intellectual battle, but merely to note that the discourse of science has, for good *and* ill, ascended to a position of cultural dominance. John Campbell's observation that "the premise of criticism in our time must be the criticism of the grammar of science" (1986, 369) is suggestive, because to a large extent, extant critical efforts have been grounded in implicit, often unreflective assumptions regarding the nature of that grammar. Hence, much of our thinking about the sociopolitical and epistemic functions of science has been hindered by a lack of conceptual clarity regarding the nature of the object of investigation, that is, of the nature of science. While there has been ample speculation about the "inputs" of science (e.g., cultural influences) and the "outputs" of science (e.g., knowledge claims or technological developments), the constitutive questions have remained hidden in the cybernetician's "black box" (Latour 1987).

Given the cultural authority accorded science and its progeny, an effort to pry open that box seems well advised. Some sort of position on the demarcation question is preliminary to the creation of the hierarchy in which that authority is bestowed. Attributed superiority assumes a prior act of differentiation. As Larry Laudan has argued, "Much of our intellectual life, and increasingly large portions of our social life, rest on the assumption that we . . . can tell the difference between science and its counterfeit" (1983, 8). Assuming that "science" and "counterfeit" are constructions rather than naturally grounded categories, a rhetorical perspective on demarcation unpacks the symbolic negotiation of that difference.

This implicit dimension of power need not be restricted to the alleged subjugation of the uninformed masses. Telling the story in this way calls into question the narrative authority of the storyteller as well. The explication of the processes in which the communal commitments of knowledge-producing communities are negotiated serves to probe internal power relations. John Nelson contends that "when the basic standards and procedures of a research enterprise come into question, we confront the political contours and contents of the research community" (1987, 429). That "confrontation" carries with it the implicit potential for critique.

One certainly does not want to call for the substantive and political "leveling" of all discourses into a sort of leftover casserole of the academy. After all, some ideas are judged to be better than others, and practitioners in certain disciplines rightly can claim to be better judges of the ideas advanced within their disciplines than other assessors. Nonetheless, scholars interested in the social constitution of science cannot afford to neglect the political implications of their claims. That is most certainly the case as regards rhetorical demarcation practices, for there is an essential social order implicated in any functional set of demarcation criteria. As Hariman has observed, "The act of comparing discourses implies both manifest definitions of substance and latent attributions of status for each genre, and the disputes about categorizing discourses often are concerned more with questions of status than of substance" (1986, 38). Indeed, as the categories of status and substance shade into one another, in a process that can best be described as rhetorical, the true complexity of demarcation practices is illuminated, and their entailments are made the objects of critical engagement—the products and objects of rhetoric.

2 PHILOSOPHICAL PERSPECTIVES

OF PROPOSITIONS, PROCEDURES, AND POLITICS

The sort of disciplinary reconstruction on which a rhetorical account of demarcation rests presupposes the local authority of its multiple prior constructions. As such, this chapter describes several dead ends and ultimately a rhetorical turn. It begins with several versions of what Holton (1981) described as Ionian enchantments, a fixation on a single, universal theory of science; and it ends with the one turn which most of the philosophers I discuss most wanted to avoid—a self-conscious move to the rhetorical.

Two observations are in order. First, my canonical history is obviously (and admittedly) partial and incomplete. I do not presume to tell philosophy's story on its own terms. My goal here is to sketch, in very broad strokes, the background against which I will later develop a rhetorical account. Second, my particular take on this disciplinary history is more than a bit opportunistic. Indeed, I am not trained as a philosopher of science and, hence, might well not provide as nuanced an account as those so trained would prefer. I advance critical judgments that are neither "self-evident" nor likely to be accepted by scholars concerned with elaborating alternative views of demarcation. I do not presume that these judgments proceed from scholarly insights necessarily superior to others advanced for other purposes, nor do I claim that the history of philosophy is but one immutable march on the road to rhetoric, where we find Rouse and Fuller "resituated at the end of philosophy," as Schrag might put it. That would be revisionist history of the highest order.[1] A rhetorical perspective on demarcation is not designed to answer the demarcation question, at least as it has traditionally been phrased in philosophical terms. Rather, it asks the question in

1. Even a philosopher purportedly aware of and sympathetic to rhetorical studies like Henry Johnstone has labeled rhetoric "philosophy without tears" (1990, xvii). A better example of the synergistic relationship of rhetoric and philosophy can be found in Schrag 1992.

different terms and then draws from the philosophical literature in its search for its own answers. As a consequence, I am interested less in detailing what philosophers "really meant" or in clearing up the confusion which remains than in a limited appropriation of philosophy for the strategic purpose of crafting a rhetorical account.

Demarcating Science as Knowledge Claims: From Propositions to Programs

Traditional philosophical speculation has long taken as one of its primary foci the relative legitimacy of various ways of knowing, situating science as an especially dominant concern. Viewed broadly, the philosophy of science has generally been concerned to elucidate the conditions under which scientists may properly claim to have produced legitimate knowledge and under which we should accept it as such (Fuller 1989b, 78).

To enter a discussion of science at all, we commit ourselves implicitly to some position on the demarcation issue. Amsterdamski has argued that "any theoretical stance on science must . . . base itself on a more or less articulated definition of science. . . . It is impossible to study science without accepting in advance some opinion concerning the scope of the phenomenon, the evolution of which one attempts to investigate" (1975, 24). Indeed, such a concern has been construed as the *most* crucial topic in contemporary discussions of human knowledge. Bartley insists, for example, that "for some time two strident motifs of our intellectual life have been the efforts of scientists and non-scientists to come to terms with one other, and the efforts of scientists to state just what it is that makes what they do scientific" (1968b, 42).

Unlike most disciplines, which have traditionally begged the intellectual question of the constitutive nature of science, philosophy has sustained a self-conscious tradition of argument on this very issue. Indeed, it would be no exaggeration to suggest that philosophers have traditionally assumed that they "are supposed to legislate the principles that scientists use as the basis of their research adjudications" (Fuller 1989b, 78). Overseeing the boundaries of rationality by arbitrating and evaluating the claims of aspirants to the mantle of the truly scientific, they erect barriers to the encroachment of the merely "pseudoscientific." As Larry Laudan suggests, "It is small wonder . . . that the question of the nature of science has loomed very large in Western philosophy. From Plato to Popper, philosophers have sought to identify those epistemic features which *demarcate* or mark off science from other sorts of belief and activity" (1983, 9).

What I want to highlight here is that this prescriptive "search for identity" has been largely unconcerned with the actual conduct of science, except insofar as it fails to live up to the standard in question and thus invites a characterization as "metaphysical," or at least as "nonscientific." Giere, for example, argues that "the scientist of philosophical theory is an ideal type, the ideally rational scientist" (1988, 2). In this sense, science is represented, perhaps with good reason, as a superior way of knowing.

Second, traditional philosophical approaches systematically exclude the human element in science: "The actions of real scientists, when they are considered at all, are measured and evaluated by how well they meet the ideal" (Giere 1988, 2). Consequently, the *context* of scientific investigation, the production and legitimation of scientific knowledge, is typically regarded as of little consequence for a proper philosophical understanding of science. As a preliminary observation, then, it seems that philosophical approaches to the demarcation question downplay the practical influences on and consequences of the various standards advanced.[2] This is not to say that philosophers must necessarily do everything, in addition to doing philosophy. It is simply to say that, given the presumed legitimacy of the philosopher's dominion over the *true* nature of science, it would seem that a broader perspective would be warranted. Amsterdamski, for example, has conceded that "regardless of the intentions and goals determining the way in which particular philosophies construct the concept of science and specify the criteria of scientific method, they are, nevertheless, in some manner, . . . 'responsible' for the vision of science functioning in sociological or historical studies and for its common sense understanding" (1975, 24; see also Fuller 1988). It is here that, ultimately, the episteme of philosophy meets the praxis of rhetoric.

Twentieth-Century Philosophy of Science on Demarcation

While discussions of demarcation are hardly unique in this regard, the dominant forms of these discussions share a common intellectual debt to logical empiricism and its twin pillars of foundationism and logicism. As Giere put it, "It had no serious rivals. Within the philoso-

2. This is not to suggest that, as a general rule, philosophers have neglected the political implications of their scholarship. My point is that, as traditionally phrased, the demarcation question has not been understood as intrinsically practical and political, assumptions at the heart of a rhetorical account.

phy of science, it still provides the primary foil for current investigations" (1988, 22). This dominant brand of empiricism would constrain, even dictate, the nature of demarcation proposals at least until the so-called Kuhnian Revolution in the 1960s.

The logical empiricist commitment to foundationism is most important insofar as it legitimates a philosophy of science without a necessary concern for the particularity of scientific *practice.* Considerations of what scientists actually do were clearly subordinate to the legitimation of science as a series of epistemic propositions. This insistence on a non-descriptive and justificatory philosophy of science would remain the philosophical norm for most of the twentieth century. Most representations of science, the construction of which were prerequisites for their legitimation, were necessarily idealized.

The related commitment to logicism, perhaps illustrated most clearly in Gottlob Frege's position that logic was not the science of thinking per se, but rather the study of purely logical relationships among objective abstract entities, exerted similar influence on discussions of demarcation. From such an orientation, as Carnap put it, "Philosophy is to be replaced by the logic of science—that is to say, by the logical analysis of the concepts and sentences of the sciences, for the logic of science is nothing other than the logical syntax of the language of science" (1937, xiii).

Clearly, such a commitment makes the philosophy of science in general and demarcation proposals in particular a priori endeavors. Its concern is with how, ideally, a scientist should go about doing the business of science, rather than, say, how she or he actually goes about it or what the implications might be of differing ways of going. This characteristic removes from consideration any semblance of the social context of scientific practice and makes "scientificness" a property not of scientists or their practices, but only of their statements.

While few, if any, contemporary philosophers of science would still defend either of these doctrines in anything resembling its original (and most restrictive) forms, this genealogy has left its mark. The first vestige of this legacy is that the products of science, that is, scientific theories, are essentially abstract linguistic structures. This appears to militate against a meaningfully social (hence, rhetorical) conception of science and scientific practice. The second residue of empiricism involves the consistent presumption that the task of the philosopher of science (particularly in demarcation discussions) is to "reveal" the pre-existent, yet unclear, rules of "truly" scientific theorizing.

Verificationism and Falsificationism

One early perspective on the demarcation question is best understood as verificationist. Articulated by Wittgenstein in his *Tractatus* (1961) and subsequently elaborated by Schlick, Carnap, and their Vienna Circle colleagues, this doctrine maintains that the meaningfulness of any statement (i.e., truly scientific postulate) is predicated on its verifiability, on its capacity to be "proved true." Hence, verificationists recognize the existence of two types of statements: "meaningful" ones, which can be proved true, and "metaphysical" ones, which cannot be so proved. This lack of provability need not entail falsity; it may simply indicate that said statements are beyond adequate proof of truth. The "truth" of meaningful statements can be demonstrated either by empirical verification or by analysis of the meanings of their components. Therein lies the proposed demarcation criterion: the statements of the empirical sciences are capable of empirical verification. Those statements in the latter category are considered the formal sciences, such as logic and mathematics.

The limitations of this early formulation are clear. At the very least, it classifies scientific theories as meaningless, hence unscientific. Insofar as they are universal statements, theories cannot be "proved true." At best, they can receive ancillary empirical confirmation. A second flaw lies in the conclusion that even the strongest confirmation at a given point in time does not and cannot exclude the possibility of future empirical disconfirmation.

Consequently, Carnap (1967) and Reichenbach (1938, 1951) replaced verifiability with the related, yet somewhat more liberal, criterion of confirmation or testability. While they maintained the distinctions between metaphysical, analytic, and empirical, they "assumed that the *scientific* status of statements (their meaningfulness) depends upon the possibility of confirmation by empirical evidence" (Amsterdamski 1975, 26). A number of conceptual difficulties surrounded confirmation (such as those involving statements regarding unobservable entities), so the standard underwent a number of "liberalizations," which produced an even more thoroughly confused and confusing criterion of demarcation (Carnap 1967).

At both levels, the fundamental weakness of the verificationist position lies in its commitment to inductivism and the concomitant belief that the status of scientific statements hinges on the degree of their inductive confirmation (or verification). As Lakatos noted, verificationists held that "only those propositions can be accepted into the body of science which either describe hard facts or are infallible inductive generalizations from them" (1971, 92).

Such a commitment establishes an extremely demanding (and transcendent) epistemic standard for the "truly scientific." If an inductivist accepts a given statement, it is accepted as demonstrably and manifestly true; if it is not, it is rejected. This epistemic rigor is strict and unforgiving. A proposition must either be proved from facts or directly derived from propositions already so proved. There is no corresponding requirement that a metaphysical statement be proved false. As Lakatos argued, "Inductivist criticism is . . . primarily skeptical: it consists in showing that a proposition is unproven, that is, pseudo-scientific, rather than in showing that it is false" (1971, 93). Regardless of the degree of confirmation required (certainty, as with early justificationists, or high probability, as with neo-confirmationists), it is clear that "science," as a series of propositions, was demarcated primarily by its epistemic rigor. In effect, all knowledge claims were construed as subject to universal standards of strict induction. Only those which passed muster could be considered epistemically secure and granted the imprimatur of "science."

The foregoing demarcation criteria are hardly congenial to a rhetorical account of demarcation, or even to most contemporary philosophical discussions. Most notably, the myopic focus on sets of propositions and artificial epistemic standards leads to an utter (and self-conscious) rejection of any role for any external influences on the production and legitimation of scientific knowledge claims. In keeping with Reichenbach's "contexts," processes of production are removed from legitimate inquiry. More to the point, however, is the assumption that sociocultural influences on the legitimation of knowledge claims are intellectually taboo. If there is evidence of some external influence on the acceptance of a given set of statements, it is argued, one must withdraw's one's acceptance: "Proof of external influences means invalidation" (Weimer 1977, 2–3; see also Hempel and Oppenheim 1948; Lakatos 1971). As a consequence, logical rigor comes at the cost of practical epistemic relevance.

The confirmationist program also underestimates the central role of individual interpretation in making evaluations of relative confirmability. There is no clear indication as to the point at which confirmation is achieved. To the extent that the designation of "scientific" directly depends on the identification of that point, such a glaring weakness would seem effectively to invalidate the confirmationist program as one which can demarcate science from other ways of knowing. Insofar as it is unclear how a true confirmationist would gain immediate and unproblematic access to a "natural world" which presumably would serve as the arbiter of confirmation, the confirmationist program is descriptively and evaluatively "meaningless."

It might well be argued that one should not be too quick to dismiss empirical reality, or certain of its abstract linguistic representations, as constraints on science and the assessment of its knowledge claims. I will not address that general issue here specifically. I would argue, however, that *any* such constraint is intrinsically interpretive and, more to the point, requires the authorization of some relevant community. That such a communal authorization might be, at a given time, unanimous on fundamental claims—for example, the earth revolves around the sun—is not evidence that such claims are scientific because they are intrinsically confirmable (or confirmed, empirically). It is evidence that a particular authorizing audience has negotiated the boundaries of "science" and communally established particular standards of "acceptable epistemic warrant" for such claims which are advanced within those boundaries.

Sir Karl Popper (1957, 1959, 1968), motivated in large part by the untenability of the inductivism underwriting the verificationist program, articulated perhaps the most famous demarcation criterion: falsificationism.[3] While he chose quite different means than the verificationist, the general ends were the same. In *The Logic of Scientific Discovery*, he sought a demarcation criterion as a solution to "the problem of finding a criterion which . . . distinguish[es] between empirical sciences on the one hand, and mathematics and logic as well as 'metaphysical' systems on the other" (1959, 34). He would subsequently include "pseudoscience" as a system to be distinguished from science as well (1982, 175).[4]

Legitimating as scientific only those theories that forbid certain observable states of affairs and therefore are factually disprovable, Popper insisted that "it must be possible for an empirical scientific system to be refuted by experience" (1959, 41). Additionally, he maintained that in order to be considered scientific, a theory must satisfy the additional requirement of predicting novel facts, that is, facts which are unexpected in the light of previous knowledge. Consequently, it goes against the falsificationists' "code of scientific honour to propose unfalsifiable the-

3. Wisdom goes so far as to suggest that "in its full articulation the demarcation problem is properly to be regarded as Popper's problem" (1987, 42). His point is that, while some attention to the nature of science is a prerequisite for saying anything about science, Popper was the first to engage the topic self-consciously as an end in its own right. While the question of paternity holds little purchase on my discussion, this seems a difficult argument to sustain.

4. Contra much popular commentary, Popper does not exclude the possibility that so-called metaphysical, even pseudoscientific (nonfalsifiable) systems could be of immense value to the scientific enterprise (Popper 1982, 185–216).

ories or 'ad hoc' hypotheses which imply no novel empirical predications" (Lakatos 1971, 97).[5]

The laudable motive of Popper's definitional claims appears to be the general enhancement of scientists' claim-making, at least in the sense of excluding propositions that elude appropriately rigorous critique, refutation, and/or falsification. It is in this sense that Popper is generally considered the exemplar of critical rationalism. For him, science is one among many forms of human myth-making. It is a form, however, which is distinct from "all others" by a second-order tradition of the critical discussion of its constituent myths (Popper 1962). For critical rationalists, persistent criticism of knowledge claims is the hallmark of scientific inquiry. The proper conduct of science, then, would involve the incessant attempt to falsify the empirical predictions or observations advanced by practicing scientists in a series of "conjectures and refutations." The key for Popper (and what would set him apart from, e.g., Kuhn [1962] and Toulmin [1972]) is that "the criteria of refutation have to be laid down beforehand: it must be agreed which observable situation, if actually observed, means that the theory is refuted" (Popper 1962, 33).

This is a problematic yet important development in the demarcation controversy. While Popper might be accused of some hasty generalizations, for instance, that nonscientific myths cannot be embedded in a critical tradition, his suggestion that the standards by which "scientificness" (falsifiability) is assessed are agreed upon implies that such standards are the products of historically situated processes of negotiation. There is some cryptic evidence for this interpretation. While expressing doubt that "methodology is an empirical science," he commented that "my doubts increase when I remember that what is to be called a 'science' and who is to be called a 'scientist' must always remain a matter of convention or decision" (1959, 52). Speaking specifically of his own program, he wrote, "My criterion of demarcation will . . . have to be regarded as *a proposal for an agreement or convention*" (37).

On the other hand, if Popper means nothing more than that science is to be decided by the application of a priori standards, then he articu-

5. While it is popularly assumed that Popper's standard is relatively simple and has remained static since its first articulation (Casti 1989), such is clearly not the case. Lakatos, for example, has argued convincingly that Popper evolved from a naive falsification which conflated nonfalsifiability with disproof of a theory to a more mature "methodological" falsificationism which maintained that only those "nonobservational propositions" which forbid certain states of affairs are scientific (Lakatos 1971). I will not engage the particulars of his transformations insofar as they do not alter his fundamental rejection of inductivism and his commitment to falsification (of whatever variety).

lates a truncated conception of criticism. While it is unclear which (if either) of these interpretations Popper would accept, it is enough to note that Popper does allow some, albeit insufficient, critical "room" for the formative influence of an authorizing community.[6]

An additional dimension of Popper's rich formulation which merits attention here is his recognition of the allied functions of explanation and argument. He contended that argumentation is impossible absent the use of language to describe: "An argument, for example, serves as an expression insofar as it is an outward symptom of some internal state. . . . Insofar as it is about something, and supports a view of some situation or state of affairs, it is descriptive" (1962, 295). He suggests that explanatory and argumentative dimensions or uses of language cannot be separated because of the "logical analysis of explanation and its relation to deduction" (135).

I agree with Popper's alliance of explanation and argument, and his understanding of the "language of science" seems exponentially more nuanced than Carnap's (1937) focus on logical syntax. While all discourse is advocative, it must be noted that Popper's alternative notion of argument is as problematic as the confirmationists' induction, although on quite different grounds. Explanation, "scientific" or otherwise, cannot be limited to the strictures of formal predication and entailment.

By presuming the potential of at least local evaluative closure, processes of argument implicate normative criteria. Hence it is important to consider the normative dimension of the falsification criterion. It seems abundantly clear that Popper sees his philosophical speculations as a means of constraining the course of "true science." He observes at one point, "I shall try to establish the rules, or if you will the norms, by which the scientist is guided when he is engaged in research or in discovery" (1959, 19). Scientists, he maintains, ought to use a set of rules that ensure that their propositions are falsifiable and that propositions are accepted only if they have withstood rigorous attempts to show them to be false. Only to the extent that these rules are instantiated can genuine scientific knowledge be generated.

In an important sense, then, Popper idealizes scientific practice insofar as he advances rules of scientific method which follow from his philosophical demarcation position, rather than from scientific prac-

6. Popper's position on the conventional (consensual?) nature of demarcation criteria is further complicated by his criticisms of those "conventionalisms" that he takes as nonempirical. He comments, "I find it quite unacceptable. Underlying it is an idea of science, of its aims and purposes, which are entirely different from mine" (Popper 1959, 82). Nola (1987) offers a cogent critique of Popper's inconsistency on the acceptability of "conventionalist stratagems" (451).

tice. The degree to which those rules are associated with science remains frustratingly ambiguous. At one point he argues that questions of whether his demarcation criterion can or should affect scientists' theory choices "are quite independent of this [demarcation] problem" (1959, 31–32). Later in the same work, however, he rejects a purely logical analysis of particular "scientific propositions" in favor of an analysis of "our manner of dealing with scientific systems . . . what we do with them and what we do to them" (50).

This tension is particularly troubling if one is at all concerned to offer sociological or rhetorical accounts of scientific demarcation practices, since it militates actively against the application of Popper's most fundamental commitment, namely that scientists should (and, by definition, must) formulate conjectures which can be clearly refuted. This is clear enough as long as one confines attention solely to the logical relations of constituent propositions. Popper recognizes, however, that the asymmetry between proof and disproof becomes highly problematic in practice: "In point of fact, no conclusive disproof of a theory can ever be produced, for it is always possible to say that the experimental results are not reliable. . . . If you insist on strict proof (or strict disproof), you will never benefit from experience, and never learn from it how wrong you are" (1959, 50).

Popper's falsificationism cannot begin to account for the "true" nature of science insofar as he recognizes that a strict application of his demarcation criterion would, by necessity, exclude all forms of presumably scientific practice. It could be argued that this is evidence that no criterion is possible and we should abandon Popper's and everyone else's search for one, or perhaps just make more pragmatic our standards for a reasonable criterion. That is an alternative to be taken up later. For current purposes, it should suffice to note that Popper's particular formulation cannot be applied in or accommodated to the vagaries of the scientific enterprise. Such a nonempirical perspective leaves falsificationism, as Nola puts it, "untrammelled by actual science" (1987, 458).[7]

The above discussion of the ambiguity of scientific practice in Popper's formulation has alluded to an issue that should be made explicit. If demarcation criteria are be to anything more than "toothless wonders" (L. Laudan 1983), they must have some relation to our understanding of the structuring and evaluation of scientific practice, though who participates in these practices remains arguable. This is not to sug-

7. However, Nola does suggest that Popper's "post-LSD" (a provocative description) makes an "attempt to link his rules of method to actual science" (1987, 469).

gest that a timeless, transcendent prescription for science is available, less yet desirable. It is to suggest that any criterion of demarcation must be normative in the sense that it is used by scientific rhetors to legitimate certain sets of practices and delegitimate others. More to the point, it is to suggest that "its acceptance or refutation is always a matter of convention" (Amsterdamski 1975, 30). Indeed, criteria for judgment exist only in and through their historically situated communal articulations.

Popper's influence on philosophy generally and on the demarcation issue in particular is undeniable. Like a modern antihero, his falsificationism offers a paradoxical combination of foils for and contributions to a rhetorical account of demarcation. His idealized view of the truly scientific as falsifiable sets of propositional claims seems to foreclose the option of meaningful rhetorical demarcation practices per se. On the other hand, his cryptic allusions to scientific status as conventional and to the argumentative processes of scientific claim-making are heuristics that are borrowed here in the service of crafting a rhetorical account of demarcation.

These heuristics of disagreement are perhaps best illustrated in the work of Popper's one-time student and most insightful commentator, Imre Lakatos. Lakatos' interpretation and modification of central Popperian themes are crucial contributions to the demarcation problem. Smart argued that Lakatos' "own positive and very original suggestions are very much within the general spirit of Popperian philosophy" (1972, 269). Lakatos does, however, go well beyond Popper's original conceptualizations.

Lakatos' perspective on the philosophy of science generally and the demarcation problem in particular is known as the "methodology of research programmes" (1970, 1971, 1978). In his own words, this "account implies a new criterion of demarcation between 'mature science,' consisting of research programmes, and 'immature science,' consisting of a mere patched up pattern of trial and error" (Lakatos 1971, 124). While Lakatos' own programmatic statement manifests less concern with the *definition* of science than with an *evaluation* of acceptable scientific programs (progressive vs. degenerative), my reading suggests that the constitutive demarcation question remains a critical issue for Lakatosian philosophy. As he puts it, "Science, after all, must be demarcated from a curiosity shop where funny local—or cosmic—oddities are collected and displayed" (1970, 102).

While the precise definition of "scientific research program" remains vague, it appears to be a *sequence* of theories in which certain methodological rules are followed. The primary components are a "hard core" of immutable hypotheses; a "protective belt" of auxiliary hypotheses; a "negative heuristic," or assumptions underlying the hard core which are not open to question; and a "positive heuristic," or a set of conditions under which the research program is to be changed (Lakatos 1970, 132ff.).

The first specific characteristic of Lakatos' program that is crucial to a rhetorical account of demarcation is that the "unit of analysis" in the demarcation question is shifted from a particular set of logically related propositions to the research program itself. He writes, "The basic unit of appraisal must not be an isolated theory or conjunction of theories but rather a research programme, with a conventionally accepted . . . 'hard core'" (Lakatos 1971, 99). It is important to observe that the allegedly immutable nature of the hard core is not taken as a "necessary" characteristic of the research program and its empirical commitments. Rather, that immutability is a matter of conventional acceptance, a function presumably of historically situated processes of negotiation. While a more specific discussion of the implications of this characteristic will follow, it is important to understand here that Lakatos, at least indirectly, "carves out a niche" for consideration of social influences on and in scientific practice.

Lakatos also offers a mechanism for the validation of his own and other general demarcation criteria. His "meta-methodological rule" maintains that if a demarcation criterion is inconsistent with the "basic" valuational assumptions of a given scientific elite, the criterion should be rejected (1971, 111–112, 120). This rule is predicated on Lakatos' observation that "while there has been little agreement concerning a *universal* criterion of scientific character of theories, there has been considerable agreement over the last two centuries concerning single achievements" (111). Lakatos maintains that a scientific elite (the nature of which unfortunately remains unclear) makes evaluations of particular claims, research program, and so on, as a matter of course. As such, the speculations of philosophers about demarcation criteria should be roughly consistent with the historically situated practices and judgments of that elite.

It is clear that Lakatos' "methodology of scientific research programmes" provides valuable detail to the present story. Initially, his formulation avoids the apriorism of its intellectual predecessors. Whereas the justificationists maintained a rigidly bivalent approach to the demarcation question and Popper's formulation was primarily an analysis of logical relationships between propositions, Lakatos' criterion allows for the consideration of temporal factors in the assessment of "relative progressiveness." While justificationist and falsificationist positions can be read as espousing "statute" law, Lakatos appears bounded by a less formulary conception of "case" law. For example, his suggestive allusion to the evaluative rhetorics of the scientific elite as grounds for adjudicating a particular program's case for progressiveness points toward the value of a rhetorical account of demarcation in explicating the discursive dynamics of those judgments.

Second, Lakatos' program warrants at least some speculation on the nature and grounds of scientific change. While a rhetorical perspective on demarcation is not centrally concerned with, for example, theory choice, it will be concerned with the constructive influence of scientific controversy. If I read Lakatos correctly, he seems to suggest that the "hard core" of a given research program is constituted by the conventionally accepted set of inviolate assumptions which form the basis of theory choice, experimental practice, and the like. This clearly suggests that those assumptions serve as resources on which scientific actors may draw either in the justification of their actions or in critique of opposing research program (Siitonen 1984).

Admittedly, Lakatos does not speculate on such a function for the assumptions contained in the "hard core." Indeed, it is likely that he would construe them as something akin to tacit knowledge which is neither articulable nor adaptable, or even a focus of conscious reflection. Nonetheless, this need not prevent a more "rhetorical" reconstruction of these assumptions. As such, a major contribution of Lakatos' program is its own "heuristic" fruitfulness as regards the guiding assumptions which constrain scientific choices and (can) serve as resources in scientific argumentation.

Lakatos' position clearly represents an improvement over its predecessors, to the extent that it allows at least some consideration of the historical and social dimensions of science. He notes that "the methodology of research programmes, like any other theory of rationality, must be supplemented by empirical-external history. No rationality theory will ever solve problems like why Mendelian genetics disappeared in Soviet Russia in the 1950's. . . . Moreover, to explain the different rates of development of different research programmes, we may need to invoke external history" (1971, 102). Clearly, then, Lakatos holds out at least some room for the influence of historicocultural factors on scientific practice, though his account remains hamstrung by insistence that real science somehow transcends its "local" character.

This is not to suggest, however, that Lakatos' conceptualization and use of history are unproblematic from a rhetorical perspective. Indeed, they often contribute to the very objectification of "the scientific" that needs rethinking. Consider his suggested division between internal and external history. While these categories may carry some validity at face value, they seem to beg precisely the question that Lakatos purportedly wants to answer: What is the line of demarcation between the scientific and the nonscientific? Lakatos thus construes the distinction between science and "society" as unnecessarily rigid. In contrast, the telos of a rhetorical account is to interrogate how those lines are drawn and redrawn in rhetorical praxis.

This problem becomes even more troubling when one considers that Lakatos steadfastly defends a program of the "rational reconstruction of history" in which "internal history is history reconstructed as if the scientist had always proceeded in a rational manner" (Smart 1972, 267; see Lakatos 1971). *Rational,* in this context, of course, is taken to mean "in accordance with appropriately scientific standards," that is, those of progressive research program. External history, on the other hand, is operationalized as "what actually happened and it will take note of such non-rational factors as the social pressures on a scientist, accidents of communication . . ." (Smart 1972, 267).

In addition to glossing the problem of interpretation in historiography, there is clearly a sense in which history is to be reconstructed in the service of science. Kuhn, for example, argued that Lakatos' historical method is "not history at all but philosophy fabricating examples" (1971, 120). While the veracity of any given historical interpretation is not the concern of this discussion (see Freudenthal 1988; McAllister 1986), Lakatos' straightforward admission that he seeks, Whiggishly, to reconstruct history *in order to* demonstrate the accuracy of a given perspective on science is striking. He notes that "what looks like a phenomenon of 'irrational' adherence to a refuted theory . . . may well be explained in terms of my methodology *internally* as a rational defense of a research programme" (1971, 102). Even if one were to grant that all historical interpretation is necessarily subjective, or at least intersubjective, it is clear that Lakatos conceives of his goal as a sort of apologia for "science." The implication is that historical factors are truly meaningful only insofar as they can be neutralized in the "rational reconstruction of history." A rhetorical reading, concerned with alternative constructions within given historical contexts, calls into question any particular reconstruction of history that masks or denies the influence of those historical contexts.

A final conceptual limitation of Lakatos' program relates to the previously described "meta-methodological rule," which decrees that demarcation criteria should be consistent with the evaluations of a reigning scientific elite or be rejected as inadequate. Initially, it seems indirectly to sanction an authoritarian conception of scientific legitimation. Given that Lakatos offers no meaningful role for the constructive influence of historical and social factors in scientific practice, the sole authority for the evaluation of scientific claims, even the demarcation of science itself, is granted to a particular, historically constituted scientific elite. Those factors that Lakatos would banish to the realm of "externality" are also crucial influences on elites in their contextual negotiations of the nature of science and, even more fundamentally, on the rhetorical grounds on which "elite" status is constructed.

This standard also begs the question of what it means to be a member of a scientific elite, as opposed to, say, a political one. Following Lakatos' own standard, a scientific elite would presumably be an elite cadre of researchers involved in a progressive research program. This, of course, raises its own points of confusion: Are we to assume that the standards a given scientific elite would apply would be consistent with Lakatos' progressiveness standard? More important, I think, is the related issue of potentially multiple elites applying alternative standards of evaluation. Given at least some historical variance among scientific evaluation practices (as assumed by the reconstruction method), the *relevant* authorizing elite is unclear.

A final limitation of the meta-methodological rule is that, simply put, it renders the philosopher's contribution to this discussion trivial. If, as Lakatos seems to suggest, any given philosophical demarcation criterion must give way to a contradictory judgment by a scientific audience, then one wonders what the philosophical criterion buys us in the first place. At the very least, it would seem that philosophers should provide a criterion which is reflective of the actual demarcation behaviors of particular scientific communities. Philosophical criteria become irrelevant at the only time they could potentially make a difference, that is, when they could tell us something about demarcation that the practices of scientists don't already tell us.

This is a subtle, but critical, shift in the way we understand the demarcation issue. Rather than taking our task to be the identification of the standards that distinguish science from, say, metaphysics or religion, we now are concerned with the acts of demarcation themselves—with the ways in which boundaries are discursively (re)drawn. Perhaps our concern, then, should be with how and why those communities make the judgments that they do, suggesting that demarcation practices are best studied as processes of argument conducted by relevant social actors (including scientists) within particular sociohistorical contexts.

In sum, Lakatos offers an important remediation of the earlier verificationist and dogmatic falsificationist perspectives on the demarcation question. He provides a rudimentary social perspective on the nature of science and rejects the stifling a priority of his predecessors. He maintains, however, a counterproductive commitment to the constructive influences of sociohistorical factors and ultimately undercuts the legitimacy of his own philosophical program.

However broadly cast, the philosophical positions surveyed to this point have dealt with the propositional characteristics of science and/or the relationships obtaining among and between such propositions.[8] While the positions may differ fundamentally, they nonetheless share a common commitment to the portrayal itself. Here I want to take up alternative portrayals of science as method.

Scientific Method and Demarcation

While portrayals of science as propositional were the dominant influences on the professional philosophical discourse, the view of science as a method exerted similar influence within that professional formation and formed the staple of secondary school science curricula.

Insofar as a simpleminded notion of "scientificness" inhering in "ideas" was intellectually passé by at least the seventeenth century, there is a sense in which the *method* of either knowledge production or its subsequent legitimation became the definitive standard by which the label of "scientific" was bestowed and maintained. The near-mythic belief in some transcendent scientific method as the producer and/or guarantor of particular ways of knowing is well documented.

In this section, my intent is not to provide anything approaching a critique of every incarnation and/or mutation of method advanced as uniquely scientific. Such a task is beyond the needs of this study and has been attempted, with some success, elsewhere (L. Laudan 1968). Rather, I first want to offer a brief overview of the development of "method talk" as the ostensible defining criterion of science. Second, I will argue that placing demarcation responsibility in "method talk" makes several untenable epistemological and historical assumptions which actively militate against a meaningful understanding of scientific practices, especially to the extent that we want to understand those practices rhetorically.

8. Nola has insisted that, while perhaps only implicit in Popper's work, there is clearly an emphasis on falsifiability as a "logico-epistemological" property of knowledge claims, as well as an emphasis on the "methodological use of the notion of falsifiability" (1987, 453). While what Popper "really meant" is not especially germane to this discussion, it seems that, in either case, the result is to legitimate particular sets of *claims* as scientific. Whether that legitimacy is attributed to their internal structure or is a product of a particular theory of method, the legitimacy of the claims remains the issue of consequence.

Wisdom (1987) suggests that the fundamental and self-conscious promotion of method as the key to the demarcation problem can be traced to Francis Bacon, who "was overtly concerned to oppose the obscure and undecidable speculations of the middle ages and replace them by modern scientific methods of enquiry" (42). In *Novum organum,* Bacon wrote that "the true method of experience . . . first lights the candle and then by means of the candle shows the way; commencing . . . with experience duly ordered . . . and from it deducing axioms, and . . . again new experiments" (1900, 1.1xxxii). It is generally assumed that Bacon placed undue confidence in this method's ability to produce claims whose scientific merit was not intrinsic, but rather resulted from their production by this method. Cohen, for example, attributed to Bacon a belief "that his method could even in the end produce conclusively certain results" (1980, 222).[9] Indeed, Bacon indicated that, whatever the propositional content of science, its "business" should be done "as if by machinery" (1900, 4.40).

Of course, this view of Bacon as the father of scientific method is at best a rough sketch, leading some to insist more strongly that it is a case of mistaken paternity (Urbach 1987). Point of origin notwithstanding, the general notion of a efficient, effective, and unitary scientific method continued to flourish in the eighteenth and nineteenth centuries. Appeals to the supposed use of method became an important means of distinguishing between scientific and nonscientific approaches to the natural world, and "there was wide agreement inside and outside the scientific community about the method which made natural science the most secure form of knowledge" (Yeo and Schuster 1986, ix). It is important to note here that *method* designates not a particular approach to solving a particular experimental problem. As used here and in philosophical conceptions of scientific method, it refers to a broader construct by which science can be demarcated and scientific claims (de)legitimated.

Larry Laudan (1982) speculates that a primary motivation for the emergence of a single, accessible, and efficacious method as a defining characteristic of science was that it resonated with the generally taken-for-granted assumption that science required, indeed mandated, internal consensus among its practitioners. "Philosophers preached that science was a consensual activity because scientists, insofar as they were rational, shaped their beliefs according to the canons of a shared scien-

9. Urbach (1987) rejects the claim that Bacon was convinced of the infallibility of his method. While his analysis is convincing, my concern here is not with the relative (in)fallibility of Bacon's method or, less yet, with Bacon's beliefs about it. My point is that, fallible or not, the method was (and, to a large extent, continues to be) viewed as constitutively scientific.

tific method or inductive logic and those canons were thought to be more than sufficient to resolve any genuine disagreement about matters of fact" (Laudan 1982, 256). As noted above, many prominent philosophers of science, such as Carnap, Reichenbach, and Popper, were primarily concerned to explicate those rules of experimental and evidential inference.

While the above should not and cannot be taken as a comprehensive discussion of the development of philosophical method talk, it suffices as a suitable overview of the motivations behind the promulgation of method-based demarcation criteria. This is not to deny that certain methodologies are better suited than others to solving particular problems or answering particular types of questions. Clearly they are. My point here is simply that characterizing a particular method as the defining characteristic of science is a conceptually ambiguous and historically inconsistent position.

Initially, it seems that the dominant motivation for the elaboration of various mutations of truly scientific method, the alleged consistency with the consensuality of science, is predicated on a fundamentally flawed assumption, that of the consensuality of science.[10] The bulk of post-Kuhnian history and philosophy of science clearly suggests that scientific change is far from cumulative, and hence the notion of consensuality in scientific belief is necessarily suspect. To the extent that a level of consensus sufficient to warrant *a* method as a demarcation criteria existed, the nature of scientific change would presumably be unidirectional and cumulative, a condition that does not obtain. As Laudan has argued, "It is for just this reason that the recent discovery that theory change in science is non-cumulative and non-convergent created such acute difficulties for methodolotors" (1982, 258).

More broadly speaking, I would contend that such methodological standards of demarcation are, in many ways, little more than a tautology. If, for example, a constitutive set of procedures for the production of knowledge did exist, then the future nature of science would be unproblematically intuitable. As Sapp argues, "The growth of science would be the result of an unfolding, its future course predicted at the moment the method was discovered" (1986, 167). The implications of this position are troubling. Most important, "Only those deviant groups that have fallen by the wayside of the correct scientific method would actually have a history. Time and social context would be relevant only to those 'non-scientific' scientists whose 'non-scientific' interests . . .

10. To suggest that the scientific community is generally in a state of dissensus is not to denigrate its intellectual authority.

created obstacles to the proper functioning of the correct scientific method" (Sapp 1986, 167). By implication, then, "correct" belief would be evaluated only by scientific criteria, while "incorrect" belief would be reduced to evaluation by "nonscientific" criteria. While all definitions are, to a degree, tautologous, the tautology intrinsic to method talk is especially troubling. It not only "black boxes" the nature of scientific practice; it equivocates correct scientific method with correct belief.

It would seem a desirable characteristic of any constitutive definition that it bear at least some relationship to the practices it purports to define. Such is not often the case in the demarcation of science. Certainly, one needs simply to observe "science in action" (Latour 1987), in any of its myriad textual forms, to conclude that the allegedly universal commitment of science and scientists to a particular and particularistic method is illusory. The availability of particular equipment, higher or lower levels of funding, and more or less severe deadline pressures significantly constrain the selection of particular scientific methods. As Barnes has noted, "Belief in the real existence of a universal 'scientific method' is the product of constant idealization; it cannot be sustained in the face of *concrete accounts* of the diversity of science. . . . scientists themselves do not possess any shared single set of conventions whether for procedure or evaluation" (1974, 45–46). I am not prepared here to argue (nor do I think I need to) that there cannot be some universal method which is best suited for the production of knowledge claims. In fact, I suppose it is possible that we have already discovered said method but, given our human intellectual frailties, just haven't realized it yet. Pleading agnosticism on that issue, I need here only to maintain that *any* constitutive conception of method must be allied with the practices that it purports to define.[11] Perhaps the earlier shift in focus would be useful on this issue as well, insofar as a concern with the discursive drawing of demarcation lines makes demarcation a practical activity—be it accomplished in the laboratory, in the classroom, or in a National Science Foundation boardroom.

In a broad sense, the very notion of a transcendent method for science appears to ignore its own social grounding. Insofar as I have alluded above to the contingencies of experimental practice which necessitate methodological adaptation (to a purist, mutation), I can now make explicit what was previously unstated. Methodologies are necessarily bound (even subordinated) to the demands inherent in particular

11. As I will argue in a subsequent chapter, we must also broaden the domain of practices that we can potentially classify as scientific.

historical and practical contexts. And very often, it seems that methodological choices are born of local contingencies, stimulated and shaped by localized concerns with specific theoretical and technical tasks.

This general contention implicates the variable of human choice in the selection (and subsequent justification) of particular methodologies. To the extent that our interpretive construals of the demands of a given practical contingency form the basis of "method choice," the universal defining quality of method talk is undermined, but its contextualized inventional authority is accentuated.

The introduction of choice into our analysis of method carries its own set of implications which actively reject a discretely constitutive conception of method. Initially, the presence of choice raises the question of the grounds on which particular choices are made in particular sets of circumstances. To the extent that we want to construe social actors (including scientists) as decision makers, we can maintain that choices are made to advance particular interests or achieve particular goals. On the most limited level, the impinging necessity or desire to complete a particular experiment can bear significantly on method choice and application. Stokes has argued that "debate [about methodology] is likely to be couched much more concretely, with general principles instantiated as particular ways of dealing with specific problems. In these circumstances, technical matters arise at the same time as methodological issues" (1986, 139).

Beyond the rather mundane (yet not trivial) domain of everyday laboratory practice, there is a wider social dimension on which methodological demarcation criteria can be evaluated (and ultimately rejected). In a broad sense, it can be read as a tool for justifying as natural the historically conditioned ascendancy of science. Yeo, arguing from a historical perspective, seeks to explicate the "rhetorical dimension of methodological discourse . . . [by] arguing that certain assumptions and statements . . . were associated with debates about the nature of science, its public image, and the relationship between its various disciplines" (1986, 260). While there are occasional infelicities with his use of *rhetoric*, Yeo's point is an important one. Alternative construals of appropriate method represent more than demarcation criteria. They function as central topoi in the construction of the public image of science and as primary bases for the hierarchical internal structure of science, for example, physics as the "paradigm" of the sciences.

While the political implications of Yeo's position are generally implicit, there is an indication that method talk can be adapted to more self-consciously political ends. If we take as plausible the cultural truism that "knowledge is power" and also grant that, for better or worse, the sciences have become the exemplars of secure knowledge, we can intuit the political implications of an ostensibly objective definition of those sciences. More to the point, however, is the argument that particular method doctrines can be and are used as rhetorical weapons against particular scientific or philosophical adversaries. Laudan (1983) has argued that, more often than not, demarcation criteria have been used as "machines de guerre" against various aspirants to epistemic privilege. Popper, for example, first formulated his falsification criterion in an attempt to demean the status of Marxism and Freudian psychoanalysis. The upcoming case study of the creationism controversy, for example, demonstrates that various demarcation criteria are wielded like Popperian and Lakatosian clubs against creationists. On the intrinsically persuasive (hence rhetorical) nature of method, Fuller is succinct: "The experimental tradition on which the ascendancy of the natural sciences is based first laid claim to the scientific exemplar by appearing to be *a more powerful form of rhetoric*" (1989b, 3; see also Shapin and Schaffer 1985).

The most explicit articulation of this perspective on scientific method appears in the work of Paul Feyerabend (1975a, 1979, 1987). He espouses an engaging brand of "epistemological anarchism," because, in his view, critical rationalism has failed both as a characterization of effective scientific practice and as a politics, a philosophy of life.[12] Feyerabend maintains that "scientific method" is typically a matter of unreflectively "*using* the basic assumptions and not *examining* them" (1979, 81). As such, Feyerabend's general position on a methodological demarcation criterion is clear: "The events, procedures, and results that constitute the sciences have no common structure" (1975a, 1). Or, in less flattering terms, "scientists don't know what they are doing" (Feyerabend 1991, 5).

In many ways, I think a rhetorical account of demarcation is served well by adherence to Feyerabend's dictum regarding the lack of common structure, so long as that contention is understood as allowing that a "negotiated" common structure can emerge in particular social and historical contexts as a local production. Indeed, Feyerabend seems to gesture in this direction, observing that science works itself out in a complex series of relationships: "The adaptation does not involve a metaphysical entity, called 'objective reality,' but real relations between people and things" (1991, 140). From a rhetorical perspective, lines of

12. His view of professional philosophy is also rather uncomplimentary. He commented that philosophers are "poets without poetic talent but not without cunning; so they created a separate subject in which emotional deprivation is an asset and lack of imagination a condition of success" (Feyerabend 1991, 495).

demarcation are precisely the practices that define those relations with profound consequences for all.

On the issue of the broader philosophical and political implications of method talk, Feyerabend's comments are pointed. In his *Farewell to Reason* (1987), he argues that "the attempt to enforce a universal truth [a universal way of finding truth] has led to disasters in the social domain and to empty formalism combined with never-to-be-fulfilled promises in the natural sciences" (61). He condemns the very notion of a universal methodological "guarantor" of human rationality, arguing, for example, that "ideas will be misused unless they have some inbuilt protection. . . . a message that helps in some circumstances may be deadly in others" (Feyerabend 1979, 68). He makes a general plea for a respect for cultural and intellectual diversity. Intellectual or methodological imperialism, just like economic or religious imperialism, he maintains, is contrary to a truly humane world community. As he puts it, "*An abstract discussion of the lives of people I do not know and with whose situation I am not familiar is not only a waste of time, it is also inhumane and impertinent*" (1991, 305).

In defense of method, it could, of course, be argued that the true universality of a particular view of method is not the crucial issue. What could be construed as vital is that "science," as a knowledge-producing entity, does achieve a remarkable degree of success in solving puzzles, and so on. Indeed, this may go unchallenged. As a successful knowledge producer, it could be argued, science has at least stumbled across a set of practices which allow the work to get done pretty darn well (H. Collins 1981c).

It is one thing to suggest that science has been a remarkably productive set of social practices. It is quite another to suggest that an even roughly consistent set of methods carries explanatory value in that regard. As a consequence, method talk might serve as one of a number of resources which may be drawn upon to offer retrospective evaluations of a particular scientific claim. That alone, however, cannot begin to justify a constitutive role for method talk in the philosophy of science.

Each of these positions has, in one way or another, articulated a self-consciously prescriptive or normative perspective on the demarcation problem. Each has also been found insufficient and/or misguided in its specification of particular a priori criteria of demarcation. In large part, I think these limitations stem from the understandable, laudable desire to avoid the doctrinaire character of earlier prescriptive demarcations, whether on empirical or critical grounds. In the next section, I want to illustrate the potential value of an intellectual and political rapprochement between the extremes of hidebound prescription and the occasional nihilism of anarchistic description.

Science as Practice: An Emerging Descriptive Alternative

This "middle way" begins with the suggestion that what has been missing is a self-conscious concern with the actual conduct of scientific practice, in all its forms. In our mythic quest to define science, we must reorient our philosophical trajectory in a different direction—one that purports to offer a descriptive account of its constitutive nature.

There is, of course, a wealth of literature that seeks to describe the nature and functions of scientific practice. The next chapter, for example, addresses sociological attempts to engage that question. What follows here might be taken as the philosophical kin of these sociological accounts. Perhaps the most fundamental similarity between descriptivist philosophy of science and the sociology of science lies in the commitment to explicate the actual sets of social practice which we have come to regard as scientific. Holton, seeking a rapprochement of philosophy and science, suggested that most scientists perceive, "right or wrong, that the messages of recent philosophers, who themselves are generally not scientists, are essentially impotent in use, and therefore may be safely neglected" (1981, 23; see also Losee 1987). The descriptive "turn" in the philosophy of science offers a closer and more careful analysis of scientific practice.

Two additional caveats are warranted here, however. First, despite its concern for the practices of science, the nonprescriptive philosophy of science remains committed to a view of science as a discretely identifiable set of social practices. Consequently, its application to a rhetorically constructive conception of science is indirect, at best. It does, however, raise important questions to which such a conception could provide an answer.

I want first to consider the pioneering work of Thomas Kuhn (1962, 1965, 1977), starting with his conceptualization of the "scientific community." He argues that the relevant scientific community for the production of knowledge is not the totality of all natural or physical scientists. Rather, he maintains that it is a decidedly more restricted collective of approximately one hundred persons with broader status. He argues that they are "practitioners of a scientific specialty. . . . they have undergone similar educational and professional initiations . . . [and] have absorbed the same technical literature and drawn many of the same lessons from it" (1962, 177). The impact of this specialized community can be recognized in Kuhn's contention that its members serve as the "producers and validators of scientific knowledge" (181) in periods of "normal science."

Such a rendering of the scientific community is relevant to the current project in that it places considerable productive and epistemic power in some sort of authorizing audience. Evaluative, hence rhetorical, judgments regarding particular knowledge claims and/or sets of practices emerge from this authorizing community. While the seemingly arbitrary selection of about one hundred persons seems at least arguable on empirical grounds, it does suggest that as a hierarchical enterprise, science is centered around the efforts of practicing scientists to "reconstruct" their research activities in a manner conducive to the positive evaluation of the relevant communal authorities.

The influence of the concept of scientific community cannot be meaningfully understood apart from the broader construct of the "paradigm," perhaps Kuhn's most recognized legacy (at times, to his own chagrin). I want here only to indicate its most general nature. In Kuhn's terms, "It stands for the entire constellation of beliefs, values, techniques, and so on shared by the members of a given community. On the other [hand], it denotes one sort of element in the constellation, the concrete puzzle-solutions" (1977, 175). This notion of the true "operative identity" of a given set of scientists and scientific practices being predicated on a core of valuational assumptions and practical exemplars is important.

Initially, it focuses attention on the substantial investment required by members of a given paradigm-defined scientific community. To the extent that membership (and presumably, the benefits of membership) are conditional on the acceptance of paradigmatic assumptions, then those assumptions can easily be construed as the foundations for negative responses to unorthodox claims which threaten that scientific community's commitments. Consequently, paradigmatic assumptions are crucial weapons in scientific controversy. The solution of revolutionary battles is not to be found in an appeal to the facts of the case, because what counts as a fact is linked inextricably to broader paradigmatic assumptions. The two case studies to follow offer different, yet consistent, illustrations of the constitutive nature of paradigmatic assumptions and their functions in the closure of scientific controversy.

A second implication of Kuhnian philosophy of science concerns its implicit rejection of method talk as a sufficient constitutive demarcation criterion. Recall that method-driven demarcation criteria necessarily assume the application of a detached, impersonal methodology in the production and/or justification of scientific claims. The all-encompassing influence of paradigmatic commitments, however, makes such an objective detachment impossible. The very idea of a constitutively scientific experiment rests on a commitment to the idea that the experimenter can be detached from the apparatus used to test a given claim, a position developed convincingly by Harry Collins (1985). Kuhn's perspective, however, entails that the experimenter, the theory being tested, and the exemplary status of the testing apparatus are all manifestations of paradigmatic commitments and expectations. Consequently, the knowledge claim produced is also a manifestation of those commitments and expectations. This is not to say that the influence of such commitments renders claims suspect. It is merely to say that the presence of conflicting paradigms, hence conflicting exemplary methodologies, rejects method talk as constitutive of science.

Clearly, then, Kuhn's philosophy of science carries important implications for any discussion of demarcation, insofar as *science* is alternately defined and understood as particular sets of shared assumptions, values, and commitments.[13] His close historical analysis of particular controversies in science also suggests the importance of these implicit demarcation criteria in the actual production and legitimation of knowledge. That is a conclusion consistent with the current project.

It should be noted, though, that Kuhn does not embrace a view of science as (re)constructed literally and wholly via revolution. While dominant paradigms may come and go, what they are paradigmatic of (science) retains an essential continuity. He speaks, for example, of the "intrinsic technicality of science" (1977, 128), and Barnes has argued that Kuhn "has tended to discourage the extension of his ideas to forms of culture other than science" (1983, 15). For example, while Kuhn characterizes the innovation/conservatism dyad as enacting an "essential tension," science retains a certain essence amid the tumult of its revolutions.

Stephen Toulmin (1961, 1972, 1974) also can be read as searching for a nonprescriptive philosophy of science. As with Kuhn, Toulmin's concerns are too wide-ranging for complete treatment here. Accordingly, I will focus very narrowly on two issues that I construe as most directly relevant to the current project.

The first concerns Toulmin's discussion of the close connection between the analytic and evaluative dimensions of constitutive criteria. He commented, "Our problem has two faces. From one point of view, it calls only for labelling or classification. . . . Yet . . . the aims and purposes by which we mark off science taxonomically . . . also imply stan-

13. It might argued that I have committed a fundamental "category" error by taking up Kuhn here in a philosophical narrative. After all, he is a historian of science, and his account is largely a historical one. His locus of professional employment is of little concern here, and his method of arguing is less salient to me than what appear to be his goals. From historical evidences, he attempts to characterize science, writ large, and hence dwells extensively in the house of philosophy.

dards for judging the scientist's achievements" (1961, 14). This, then, forces the demarcation question from the insular domain of prescription into the messier domain of daily practice (Toulmin 1960, 11). As a consequence, Toulmin is openly critical of traditional philosophical "portmanteau characterizations of science," arguing that "there is no universal recipe for all science" (1961, 15).

This is not to say that Toulmin articulated a radically open-ended account of science. He called simply for closer attention to how constitutive issues are managed in scientific practice. In *Foresight and Understanding* (1961, 44–82), Toulmin considered "ideals of natural order" and emphasized the role of these basic assumptions in the development of science. Ideals of natural order are presumptive standards of regularity which "mark off for us those happenings in the world around us which do require explanation by contrasting them with 'the natural course of events'—i.e., those events which do not" (1961, 79; see Toulmin 1965).

Strongly constraining scientists' expectations of appropriate practice, these "ideals" function as reasons scientists advance to explain or justify viewing a particular phenomenon as anomalous or as (dis)confirming a given theoretical postulate. It is crucial to note that Toulmin does not offer a constitutively prescriptive analysis of these "ideals." Rather, he develops an evolutionary model to account for their *historical* development. He contends that the "identity-through-change" of a scientific discipline is analogous to the "identity-through-change" of a biological species (1972, 121–23, 135–44). According to Toulmin, conceptual development within and among scientific disciplines is a process of "natural selection" (1965).

Toulmin conceded some ambivalence on this position, writing that "[some will want to ask] can we not at least abstract certain . . . universal criteria . . . from the diversity of historical contexts and so create a secure base from which we can pass philosophical judgment . . . from outside the chances of the historical process?" (1974, 402–3). He answered with an unequivocal no, insisting that the "attempt to formulate inviolable standards for science is misguided. . . . Standards held to be inviolable at one point are qualified or abandoned subsequently. . . . It is both futile and pretentious to proscribe context-independent criteria" (Losee 1987, 125).

While the general tenor of Toulmin's position is quite consistent with the philosophical assumptions of this project, his discussion of "ideals of natural order" carries specific implications for its conduct. Initially, I want to argue that these "ideals" can be seen as strategic resources which are deployed in the initial selection and post hoc justification of particular scientific practices. As particular patterns of deployment crystallize into common configurations, they come to function as rhetorical markers for identifying "science" and its myriad "others."

While Kuhn and Toulmin offer substantial support for a nonprescriptive philosophy of science, the most self-conscious and complete articulation of such a position can be found in the work of physicist qua philosopher Gerald Holton (1978, 1981, 1988). In his chronicle of the development of the "non-prescriptive turn," Losee argues that it is in the "case studies of Gerald Holton that the full potential of descriptive philosophy of science is achieved" (1987, 135).

In the most general sense, Holton has explicated methodological and evaluative assumptions that serve as a backdrop for scientific practice within historical contexts and has traced the continuities in such practices over time. He takes as his conceptual problematic the readily apparent heterogeneity of science and scientific practice. He queries: "If scientific discourse is directed entirely by the dictates of logic and empirical findings, why is science not one great totalitarian engine, taking everyone relentlessly to the same inevitable goal?" (1981, 4).

He goes to great lengths to delimit his sphere of investigation to the public, institutional dimension of scientific activity. For Holton, the "nascent" moments of scientific production, those in which an individual practitioner makes particular decisions, cannot be treated in a truly explanatory manner except by a "psychology of discovery." Additionally, he suggests that the epistemic privilege granted science is a function, not of individual initiative, but of the public and institutional structures of science. This recognition is one he attributes even to scientists themselves: "A scientist whose external justification and approbation comes from [the institutional dimension] has little reason to self-consciously examine the nascent moment. . . . He is likely to adopt in all his discussions of science, the vocabulary and attitude of [institutional science], dry-cleaned of the personal elements" (1988, 19).

Following a method self-described as "akin to that of a folklorist or anthropologist" (1988, 17), Holton builds a theoretical model emphasizing centrality of "thematic principles" that express the basic commitments of scientists to preferred modes of inquiry and explanation. They include methodological prescriptions, evaluative standards, and high-level substantive hypotheses (1981, 16). Perhaps the most crucial of these thematic principles is the "Ionian enchantment," which represents the ideal of a unified interpretation of all natural phenomena under a small number of fundamental laws (17–23). Other such themata include "methodological standards such as 'seek within natural phenomena quantities that are conserved, maximized, or minimized'; evaluative standards such as parsimony, symmetry, and incorporation;

and high-level hypotheses such as the constancy of the velocity of light, the discreteness of electric charge, and the quantization of energy" (1988, 28–29).

Holton emphasizes that such themata may exist fundamentally in dialectical pairs of opposition. He notes that "the presuppositions pervading the work of scientists have long included also the thematic couples of constancy and change, experience and symbolic formalism, complexity and simplicity" (1988, 29). While any such quantification of principles is bound to be something of a tenuous generalization, he suggests that "a total of fewer than fifty couples seem historically to have sufficed for *negotiating* the great variety of discoveries" (30).

Two other dimensions of Holton's program merit explication. First, thematic principles are subject to change and alternative application in alternative historical and practical contexts. Certain once-dominant themes have subsequently been discarded or modified, for example, with the introduction of quantum physics. Others, such as Bohr's principle of complementarity, have emerged in the course of the development of "new sciences" (Holton 1988, 24–25). Second, Holton makes no prescriptive recommendations on behalf of the themata that he identifies. He does not claim, for example, that scientists should evaluate competing theories on the basis of simplicity rather than comprehensiveness, offering instead a description of extant evaluative procedures. The "fewer than fifty couples" cited above, however, suggest that he does not rule out the possibility of a core of themata which are more stable and enduring than others.

The implications of Holton's thematic analysis for the current project are important. Insofar as the constitutive nature of science is portrayed as a function of its identity through time instantiated by participants and derives from the shared set of thematic principles, demarcation is vested implicitly with a historically contingent nature. The constructive variability of this formulation is highlighted when Holton observes that "on this model we can understand why scientists need not hold substantially the same set of beliefs. . . . Their beliefs have considerable fine structure; and within that structure there is generally sufficient stabilizing thematic overlap and agreement and sufficient warrant for intellectual freedom that can express itself in thematic disagreements" (1981, 13).

This is a crucial shift insofar as it allows for the admittedly constructive and productive aspects of scientific practice to "fit" with a rhetorically constructive conception of science. A common objection to any such quasirelativistic program is that it forces the history of science to be an utterly random activity. Such need not be the case. Suggesting that science, as a social practice, is rhetorically constituted implies as well that it is so constituted through adherence to particular commitments, whether construed as thematic principles or topoi. The rhetorical authority of those thematic topoi resides in their ability to warrant particular experimental practices and to function as sources of post hoc legitimation for them.

Holton's position, however, is not one that I want to endorse in all its dimensions. Initially, it is not clear that Holton has provided any consideration of the actual processes in and by which the noted dialectical pairs are actually negotiated in social practice. This is perhaps not a damning criticism for one who purports to do philosophy of science, insofar as such a rhetorical analysis is typically considered to be outside traditional philosophical concerns. Those processes, inherently rhetorical in nature, are the focus of a rhetorical account of demarcation.

A related criticism involves Holton's failure to provide any direction on the social grounding of his themata. To the extent that he insists that such themata are not intrinsic to scientific practice, it seems plausible to assume that they are social constructions. Unfortunately, Holton's analysis is silent on the nature of this social construction. For example, we are left uncertain as to why any particular principle would be implicated in a given controversy or how it would be deployed.

This lacuna is more troubling in the wake of Holton's recent proclamations regarding the dangers of antiscientific beliefs, what he calls "the Beast that slumbers below" (1993, 184). Alarmed by constructivist developments in science studies as well as New Age mysticism, creationism, and similar movements, Holton shifts his telos from empirical description to outright (but relatively unexamined) prescription, pondering, "What are the earmarks of good science?" (1993, ix). This is not an unimportant question. As articulated here, however, it becomes something of a misguided attempt to invoke, rather than argue for, both a particular view of science and its value.

It has been argued that the nonprescriptive turn in the philosophy of science represents an important theoretical advance over traditional philosophical approaches. It has begun to remove the transcendent "epistemological baggage" which hampered the prescriptive tradition, has sought a clearer alliance with the social practice of science, has construed scientific assumptions as rhetorical resources, and has given indirect conceptual support for a perspective based on a social construction of science. The current project represents an attempt to build on these advancements.

Politics, Philosophy, and the Demarcation Game

Largely *via negativa,* Feyerabend and Holton implicated questions of the larger social and political significance of our differing conceptions of science. In Feyerabend's case, this served predominantly to debunk the pretensions of scientific omnipotence, while it serves for Holton as an intellectual panacea for New Ageism and related embodiments of the postmodern condition. This turn to the political is made even more explicit in recent work by Joseph Rouse, Larry Laudan, and Steve Fuller. Rouse makes this connection largely on the basis of his view of science as practice, a view in which demarcation questions lurk about the periphery. Laudan and Fuller, on the other hand, offer quite direct and self-conscious accounts of the politics of demarcation, occasionally (especially in the case of Fuller) in ways that promise a refiguration of what it means to do philosophy.

Drawing from such intellectual forebears as Heidegger, Dreyfus, Habermas, Rorty, and Foucault, Rouse (1987) rejects traditional "theoreticist" accounts of science in favor of one that assumes the inescapable practicality of scientists' labor. That labor, drawing on craftlike skills, makes science "a way [or ways] of manipulating and intervening" (38) in variously constructed "micro-worlds" (102). As a consequence, the fundamental context in which science is said to get done is the nexus of laboratory settings in which those microworlds are constructed. As Rouse puts it, the "irreducibly local character of scientific knowledge" demonstrates that "the clearly structured data that result [when the work succeeds] always reflect the local, contingent, idiosyncratic circumstances of their production" (111). Only occasionally Whiggish, Rouse maintains that "the laboratory, then, represents a mature form, not the origin of the gradually emerging project of remaking the world to make it knowable. . . . The ensemble of practices, skills, and equipment that come together in laboratories gives these older activities a new sense" (229).

Starting from this local vantage point, Rouse adds a valuable explicit political bite to his philosophical position. He maintains, for example, that "the effects of these practices upon us and our form of life . . . need to be understood in terms of power and . . . call for an explicitly political interpretation and criticism" (1987, 247).

This allied focus (on the practical and the political), while certainly a much-needed palliative to the romanticized prescriptions of science as disembodied *theoria,* seems unfortunately to undercut Rouse's pursuit of a truly reformative philosophy of science. My fear on this account is based on what I take to be the implicit distinction that is drawn between the "juridical" politics of science and "what scientists do when they *do science* rather than acting like politicians" (1987, 210). What concerns me here is that failing to problematize precisely this distinction via a close account of the politics of demarcation might well blind us to the myriad contexts in which one cannot, as a matter of practice, function as a scientist without previously, indeed simultaneously, being what Rouse would like to consider a politician.

In fairness, Rouse recognizes the potential interactions of local scientific practices and the avowedly political activities of, for example, "lobbying, testifying, and formulating and administering policy" (1987, 210). He suggests in a more recent work that "it is important to recognize that the traffic across the boundaries erected between science and society is always two-way" (1993, 13). Nonetheless, his general orientation still appears to enact an internalist dynamic, arguing that while we should probe such interactions, "we obviously cannot do this . . . until we learn to recognize the kinds of power relations in science" (1987, 210). In short, we cannot read the scientific into the social unless and until we develop more adequate vocabularies for reading the social into the scientific, a position that elides the politically significant practice of establishing the parameters and primacy of those categories to start with.

What this seems to entail, however, is an unnecessarily narrow view of both the scientific and the social. While Rouse speaks of the web of power relations as being played out in the "loosely coupled system" of science, it seems that the various constituents of that system are related causally in only one direction—from the scientific to the social. This is paradoxical when read against Rouse's powerful defense of a view of power as "shap[ing] and constrain[ing] the field of possible actions of persons within some specific social context," a view that "considerably broadens what counts as political" (1987, 211).

I do not want to suggest that all such constituents exist in equal relations of power with one another, or even that the nature of constituents or the potential range of relations can be known prior to its enactments in practice. Rather, I want to suggest that vesting "external" constituents with juridical power while seeking another sort for internal practices obscures potentially significant interactions *before the fact*. While this is something of an chicken/egg argument, I'm not sure that we need start with internalist practices as we chart the multiple power relations in which we are all bound in this technological age. A more self-conscious account of the political practices of demarcation might, then, lead Rouse's emancipatory philosophy of science to fruition.

Laudan, offering a broad historical overview of philosophical attempts to resolve the demarcation problem, argues (rightly) that such attempts have been ill fated: "It seems pretty clear to many of us . . . that philosophy has largely failed to deliver the relevant goods. Whatever the specific strengths and deficiencies of certain well-known efforts at demarcation, it can be said fairly uncontroversially that there is no demarcation line . . . which would win assent from a majority" (1983, 9).

His construal of the demands of any potential demarcation criterion is certainly grounded in the social practices of science. He insists that "any proposed dividing line between science and non-science would have to be (at least in part) explicative and thus sensitive to existing patterns of usage" (Laudan 1983, 17). While this is not clearly developed in Laudan's analysis, it does seem that he conceives of science as a set of practices which have a relatively unproblematic social identity. It is the broader constitutive question which is troubling for him. Consider his claim that "the quest for a latter-day demarcation criterion is . . . an attempt to render explicit those shared but largely implicit sorting mechanisms whereby most of us can agree about paradigmatic cases of the scientific and the non-scientific" (18). It probably can be demonstrated that there *is* a fair degree of agreement about exemplary cases at particular historical moments. The grounds of that agreement, however, remain opaque.

Another important dimension of Laudan's discussion is his clear recognition of the political dimensions of demarcation behavior. From the perspective of an "internal" history of science, he argues that various criteria have often been used by philosophers or scientists less for analytic and more for combative purposes. He cites, for example, Popper's primary concern with the exclusion of psychoanalysis and Marxism as an indicator of the "vested interests" which can underlie demarcation criteria. He further argues that "no one can look at the history of debates between scientists and 'pseudoscientists' without realizing that demarcation criteria are typically used as *machines de guerre* in a polemical battle between rival camps" (Laudan 1983, 20).

In one sense, Laudan's politicized view of demarcation is a salutary development insofar as the contemporary political implications are made a crucial topic for consideration. He argues that "the value-loaded character of the term 'science' (and its cognates) in our culture should make us realize that the labelling of a certain activity as 'scientific' or 'unscientific' has social and political ramifications which go well beyond the taxonomic task of sorting beliefs into two piles" (1983, 21). Clearly, Laudan recognizes the broader social implications of the demarcation question and insists that any proposed demarcation criterion be more than a "mere intellectual exercise."

Given his insightful comments on the demarcation problem, I find it somewhat surprising that Laudan offers a profoundly limiting solution: "The problem of demarcation—the very problem that Popper labelled the central problem of epistemology—is spurious. . . . If we would stand up and be counted on the side of reason, we ought to drop terms like 'pseudoscience' and 'unscientific' from our vocabulary; they are just hollow phrases which do only emotive work for us" (1983, 28–29). In effect, he contends that we should abandon the demarcation problem completely and content ourselves with simply seeking to determine which among our knowledge claims are epistemically warranted.[14]

This seems a false dilemma. These endeavors can be accomplished simultaneously; searching for epistemic warrant need not depend on the demarcation question and vice versa. Second, Laudan's position is unduly pessimistic, in large part because it emerges from his own commitment to the primacy of philosophy in such matters. I'm perfectly willing to grant that philosophically transcendent demarcation criteria are impossible to sustain. However, to say that the "demarcation game" should not be played by philosophers' rules need not entail that it should not be played at all. It is a game which can be (indeed is) played in practical, rhetorical terms every day. Publication or convention presentation verdicts, funding decisions, and so on are all clearly based on demarcation criteria. As such, the demarcation question has not lost its primacy. It is simply that most philosophical attempts to answer it have lost their utility. What is called for is an analysis of the actual social practices of demarcation used by scientists and other social actors to structure their methodological and epistemic judgments. To suggest

14. This is a position strikingly similar to one defended by Arthur Fine in *The Shaky Game: Einstein, Realism, and the Quantum Theory*. Fine argues that adopting "a stance neither realist nor antirealist" (1986, 5), which he calls the "natural ontological attitude," entails opting out of the "various global games, including the game of demarcation" (10). Such an epistemic stance would be "at odds with the temperament that looks for definite boundaries demarcating science from pseudoscience, or that is inclined to award the title 'scientific' like a blue ribbon on a prize goat." Rather than importing or concocting such analytic frameworks, Fine suggests, we should "try to take science on its own terms, and try not to read things into science" (149).

I share Fine's abiding mistrust of timeless, ahistorical standards of demarcation and of analytic frameworks that attempt to *reconstruct* science to meet their own assumptions. However, I'm puzzled as to how science might speak for itself. It is one thing to suggest that analyses of science should not attempt to substitute their analytic foundationalism for sterile scientism. It's quite another to suggest that science, as a series of cultural practices, comes complete with its own interpretive repertoire. The facts (of science or anything else) do not speak for themselves; they must be spoken for.

otherwise, I think, sacrifices the reformative potential of alternative demarcations at the moment in which that potential could be realized—when inserted into the public politics of demarcation.

The work of Fuller (1988, 1989a, 1989b, 1993) represents a logical place to conclude this philosophical narrative, because it simultaneously seeks to carve out a niche for traditional philosophical concerns with rationality (albeit from a decidedly nontraditional perspective) as well as meaningful social grounding for epistemic concerns. While I will not attempt here to provide a complete account of his "social epistemology" (1988), I will argue that the theoretical commitments underlying that position are important to our concern with a rhetorically constructive view of science.

Fuller's social epistemology represents an important break from earlier epistemology and philosophy of science. He recognizes two fallacies in those traditions. Initially, he contends that traditional "philosophers treat the various knowledge states and processes as properties of individuals operating in a social vacuum. They often seem to think that any correct account of individual knowledge can be, ipso facto, generalized as the correct account of social knowledge" (1988, xii). This equivocation Fuller calls the fallacy of composition. On the other hand, traditional philosophers "can frequently slip into committing the *fallacy of division* by assuming that a feature of the knowledge enterprise that appears primarily at the level of social interaction is, ipso facto, reproduced as a feature of the minds of the individuals engaged in that interaction" (1988, xiii).

The explanation for these fallacies, as Fuller sees it, lies in a fundamental underestimation of the "influence exercised by each member's expectations about what is appropriate to assert in his cognitive community, as well as each member's willingness to discount his own personal beliefs and conform to these canonical expectations" (1988, xiii). This commitment clearly situates discussions of the nature of science and its consequent epistemic judgments in social processes of negotiation in which actors advance interests (their own or communal ones) as a means of legitimation.

Whether epistemic conformity is real or apparent is of little consequence to Fuller. On the surface, it is clear that a great deal of unanimity exists in epistemic judgments. However, it is at the deeper level of scientific practice that a great deal of diversity is present. This "problematic" allows Fuller to advance perhaps the most central tenet of his social epistemology, at least from the standpoint of the current project. He contends that "because of the ease with which it can conceal epistemic differences, the *communicative process itself is the main source of cognitive change*" (1988, xiii). If we take the communicative process as the central locus of cognitive change, a rhetorical perspective on the construction of science seems eminently desirable, if only because it promises to detail more explicitly the actual processes in which cognitive change regarding the nature and evaluation of scientific claims is actually negotiated.

While sharing Laudan's disenchantment with traditional approaches to demarcation, Fuller offers a different way out of the philosophical impasse. He notes that the demarcation question is a crucial one insofar as "demarcation criteria provide an institutional means for achieving cognitive economy, which is, in turn, a necessary—albeit fallible—condition for granting epistemic warrant" (1988, 176). He maintains that, as a matter of social practice, we should require a sorting mechanism by which to evaluate the multitude of competing knowledge claims with which we are confronted. "Bracketing out" the question of whether scientific claims are actually more secure, he contends that the social role of science has, as a matter of tradition, been granted epistemic privilege. As such, "although Laudan may be right that there is no epistemically privileged way of conferring epistemic privilege, it does not follow that there is no *non-epistemically* privileged way" of conferring that privilege (Fuller 1989b, 3).

By implication, then, Fuller reconstrues demarcation criteria as social practices which are negotiated by and for those actors playing the socially sanctioned role of scientists. While the lines that these actors utter might change in fundamental sorts of ways, the social role itself remains relatively constant. This is a crucial realization for a truly rhetorical conception of science, in that it might be taken as foregrounding the social practices that are enacted rhetorically as definitive of science.

The apparent continuity of science need not prove problematic for this social construction of science. Fuller argues, consistent with his social role analysis, that (justifiably or not) the social role of science has been *granted* (earned or otherwise) epistemic authority. He suggests that "the thread that connects the history of science from the Greeks to the present day is that people come to be convinced that particular forms of knowledge are embodied in the world and are, in that sense, the sources of power over the world" (1989b, 5).

In sum, then, science, as a social practice, "does not have an essence, but is rather simply the sum of disparate strands of society that are mutually reinforced in specific places, both by the behavior of scientists and by our learned perceptual biases" (1989b, 12). More specifically, the term *science* designates not a simple practice clearly demarcated from other social practices, but a series of behaviors that combine with other behaviors to form other social practices. Science, for Fuller and this project, is foremost a profoundly social construct which may be alternately constructed but is most generally granted a high degree of epistemic authority.

The implications of this disarmingly simple conclusion are weighty. Initially, it demands a broader orientation toward the study of science, whether by philosophers or other scholars. A preliminary condition for this reorientation requires those of us in science studies to "stop thinking about science as having a natural integrity that compels the observer to interpret it exclusively in its own terms" (Fuller 1992, 416). Indeed, Fuller maintains that "the very attempt . . . to capture a culture 'in its own terms' may reflect a trenchant ethnocentric bias, which illicitly projects onto the natural attitude of the members of the alien culture" (1989b, 13).[15] The story of science, then, does not come complete with narrative closure; that is, it does not come with the self-evidently correct interpretive language.

From this interpretive commitment follows the important implication that a multiplicity of critical and analytic methods can (and should) be brought to bear on the discourses of science. The localized site of experimental practice represents a point of origin rather than the terminus for our investigation. We need not be limited to traditionally restricted research issues. In a paean to intellectual pluralism, Fuller insists that "the fact that science transpires in a given set of labs and on a certain schedule . . . alerts the observer to where and when to *start* examining how the knowledge production process works" (1989a, 31). As such, the localized nature of specific acts of production are not the sum of topics for science studies. Indeed, they may not even be the most important or interesting ones. The conception of science as a complex series of interconnected social practices implicit in a rhetorical perspective on demarcation attempts to enact the interpretive heterogeneity to which Fuller alludes.

Fuller's social epistemology and its kindred practice, knowledge policy, are developed most fully in his *Philosophy, Rhetoric, and the End of Knowledge* (1993). Urging the jettisoning of such traditional oppositions as "basic/applied," "internal/external," and even "science/society," he develops a democratic view of "prolescience" in which "knowledge

15. Given this difficulty, interpretive closure is admittedly a "shaky game" in its own right. The preferred (albeit imperfect) alternative would seem to be to subject the claims of advocates (be they scientists, philosophers, or rhetorical critics) not to the formal logics thought to be inherent in science, but to the contextually variant rhetorical logics which structure and inform deliberations within and among disparate discourse communities.

production should proceed only insofar as public participation is possible" (xviii).

Such a position might strike some (e.g., Holton) as a call for the unwarranted intrusion of the uninformed into matters best left to the relatively better informed, a condition that Bell would characterize as "impure science." I think such a reaction unwarranted. Initially, it presumes that we (and they, assuming we want to keep that polarity) can know *in advance* what constitutes the requisite knowledge to justify classification as *informed.*

Beyond this, though, it also trades on the too-seldom-examined premise that science is best organized and enacted by scientists, brooking only token opposition from constituent audiences. It seems to me, however, that even the mildest commitment to a democratic politics entails rejecting such condescension (once recognized as such) on its face. Enacting potential reconfigurations of this relationship might profitably invoke what Fuller calls the principle of "epistemic fungibility," the assurance that a given discourse's "fungibility increases with an increase in the demand that discourse places on the cognitive and material resources of society" (1993, 295), as a way of coming to judgment. In effect, then, the implication of social epistemology is, in Fuller's terms, to "take seriously the proposition that theorizing is a politically significant practice" (xi).

This book seeks to contribute to this broadened investigation of the nature and functions of science in contemporary culture. My analysis of traditional philosophical attempts to engage this topic by way of the demarcation problem suggests that those ways are incompatible with a meaningful rhetoric of demarcation. As the "descriptive turn" suggests, a more intellectually fruitful approach is to examine the historically negotiated social practices which contextually define science and its relationships with other sets of social practices. The moral of this story, however, seems to be that there are many more stories that can and need to be told.

3 HISTORICAL AND SOCIAL STUDIES OF SCIENCE AND THEIR RHETORICAL RECONSTRUCTIONS

Recently, reflecting on the past quarter century of the history of science, Nicholas Jardine concluded that "the researches of sociologists and social historians of the sciences reveal ever more clearly the sheer diversity of the ways in which practitioners of the various sciences prosecute their inquiries, prepare their findings for publication, negotiate their alliances, wage their controversies, and promote their disciplines" (1989, 15). Science, he suggests, belongs not to the realm of method or abstract propositions, but to scientists; after all, it is their inquiries that are prosecuted and their disciplines that are promoted.

I think, in many ways, Jardine is quite correct. Most notably, he calls attention to the disciplinary concern with science—as it is conducted. In this chapter, I want to extend my call for a rhetorical account of demarcation through a reconstruction of the literatures that Jardine valorizes. In so doing, I again engage in "strategic appropriation," intending not to provide an exhaustive account but to use historical and sociological insights to enrich my rhetorical account while indicating how that account could equally enrich the history and sociology of science.

The Sociology of Science: Formative Themes

Perhaps the most important figure in North American sociology of science is Robert Merton (1968, 1973, 1976). His work served as the foundation for an entire generation of research and as the theoretical foil for several others. Randall Collins and Sal Restivo remarked that "the appearance of the Merton (1973) collection underscored the fact that he had given the field its first major paradigm" (1983, 193). That paradigm was captured succinctly in his early work on the sociology of knowledge: "The discipline . . . is primarily concerned with the relations between knowledge and other existential factors in the society or culture" (Merton 1973, 366). In Gaston's broader terms, "Mertonian sociology of science is based on the idea that the institution of science can be studied as any other social institution" (1978, 2). Merton's work has been both wide-ranging and influential, and I will not attempt any exhaustive review of his research corpus. Indeed, the Mertonian program's utility in the present context is primarily as sociological material for rhetorical reconceptualization.

From an outsider's perspective, the functionalist paradigm's major impact is to highlight those social structures of the scientific community that facilitate the actual functioning of that community. Generally presumed, it seems, is that the scientific community functions *well*. Functionalist sociologists have consistently defended science against charges of irrationality, inefficiency, randomness, unfairness, and so on. Indeed, such an apologetic impulse has led to a great deal of creative reinterpretation of scientific practice, so as to provide a functionalist or efficiency-based explanation for particular behaviors. Collins and Restivo note that Merton "wanted to show that behavior that could be considered selfish and individualistic, priority struggles for example, represented a disinterested defense by the institution of science of the value it assigns to originality" (1983, 188).

Evidence of this motivation can be read in much of Merton's work as well as that of his students (S. Cole 1970; Cole and Cole 1973; Zuckerman 1977). In a 1968 essay on the "Matthew effect" (the presumptive preference for knowledge claims of established over neophyte scientists), for example, Merton argued that any unfairness on an individual level is more than balanced out on the communal level by the advantage of promoting new discoveries.[1] He wrote, "Considered in its implications for the communication system, the Matthew effect . . . may operate to heighten the visibility of new scientific communications" (1973, 447). Similarly, Cole and Cole (1973) have defended the peer review system against charges of political influence and inefficiency. Responding to increasing criticism of sexual inequality in the scientific hierarchy, Stephen Cole (1970) argued that "scientific merit," rather than institutionalized sexism, explained the imbalances ensuring male scientists myriad advantages over their female colleagues.

In keeping with Merton's dominant urge toward explicating "how well science does what it does," the individual scientist is reduced ei-

1. The particular label for the effect comes from the Bible. Matthew 25.29 reads, "But from him that hath not shall be taken away even that which he hath" (KJV). It is important to note that in the 1973 republication of the essay, Merton indicates that the contributions of Harriet Zuckerman actually warranted "joint authorship" (1973, 439n1).

ther to a building block of the broader structure or, if a potential counterexample, a pollutant of its normal "purity." In that sense, Mertonian sociology of science privileges a monolithic, albeit historically variable, conception of the scientific community and, by implication, of scientific practice.

Perhaps the most frequently discussed dimension of the Mertonian corpus is its account of the normative structure of the scientific community, the so-called ethos of science.[2] Describing the scientific ethos as "that affectively toned complex of values and norms which is held to be binding on the man of science . . . expressed in the form of prescriptions, proscriptions, preferences, and permissions" (1973, 268–69), Merton suggested that modern science has coalesced around the related norms of universalism, disinterestedness, organized skepticism, and intellectual communism. Universalism entails that "the acceptance or rejection of claims . . . is not to depend on . . . personal or social attributes" and "finds immediate expression in the canon that truth claims . . . are to be subjected to *preestablished impersonal criteria*" (270). Disinterestedness denotes a "distinctive pattern of social control" which eliminates personal bias in the production of knowledge claims. Organized skepticism requires "the temporary suspension of judgment and the detached scrutiny of beliefs in terms of empirical and logical criteria" (277). Finally, *communism* describes an "extended sense of common ownership of goods," holding that the "findings of science are a product of social collaboration and are assigned to the community" (273). In short, scientists are said to understand themselves as more or less interchangeable constituents of a community which openly and universally shares information produced by unbiased researchers who were guided by skepticism in the absence of adequate empirical evidence.[3]

As with most programmatic formulations, Merton's original formulation has been supplemented and otherwise modified in subsequent

2. In his editorial comments in the 1973 Merton collection, Storer calls specific attention to the original 1938 paper dealing with this topic, labeling it "one of the most significant in the history of science" (in Merton 1973, 226). He laments the brevity of the discussion, suggesting that it subsequently had encouraged inaccurate and unfair criticisms of the norms. Merton did offer additional comments on the norms. In any case, the frequency of positive citation makes clear the salutary impact of Merton's formulation. See, for example, Hagstrom 1965; and Stehr 1978.

3. In his analysis of priority claims in cases of scientific discovery, Merton discussed two additional values, "originality" and "humility" (1973, 292–304). However, it is not clear to me that they depart fundamentally from his earlier description; indeed, they may be comfortably incorporated into that discussion. For example, humility would seem to be the product of a truly disinterested researcher insofar as she or he disavows the personal trappings of success.

work. Barber (1952, esp. chap. 4) added the norms of individualism, rationality, and emotional neutrality, while several of Merton's students refined the initial discussion of norms to reflect a more sophisticated distinction between technical and "moral" prescriptions for scientific practice (S. Cole 1970; Cole and Cole 1973; Zuckerman 1977). Zuckerman offers perhaps the clearest summation of what now is taken as the neo-Mertonian ethos of science, arguing that it "is comprised of two classes of norms, intertwined in practice, but analytically separable: the cognitive norms and methodological canons which specify what should be studied and how, and the moral norms . . . concerning the attitudes and behavior of scientists in relation to one another and their research" (1977, 87).

The most relevant implication of this construct is that of a normative system of social control for the scientific community. In her discussion of deviance in science, Zuckerman argues, for example, that "social control in science depends partly on scientists' internalizing moral and cognitive norms in the course of their professional socialization and partly on social mechanisms for the detection of deviant behavior and the exercise of sanctions when it is detected" (1977, 90).

Read in these terms, the normative structure might profitably be understood as a powerful rhetorical vocabulary for characterizing and judging scientific procedures and propositions (Post 1986; Prelli 1989a). Consider, for example, the influence of the internalized norm of organized skepticism. To the extent that practicing scientists accept its epistemic prescriptions, experimental practice will be structured in such a way as to allow for anticipatory adaptations to counterarguments. As Merton writes, "Science is public and not private knowledge; and although the idea of 'other persons' is not employed explicitly in science, it is always tacitly involved" (1973, 219). "Other persons," of course, denotes that audience on whom the symbolic authority of the ethical vocabulary might be brought to bear, an intrinsically rhetorical process.

Similarly, the technical norm of replicability constrains the design and execution of scientific practice, as well as functioning as a post facto evaluative topos. Zuckerman maintains that "the requirement of reproducibility therefore serves not only to deter departures from cognitive and moral norms, but also makes for the detection of error and deviance" (1977, 93). Insofar as particular technical practices (e.g., replicable experiments) are considered to be partially constitutive of the nature of science, their deployment in particular controversies can be read as implicating demarcation questions. As Cloitre and Shinn put it, "The Mertonian model affirms that science and non-science are clearly demarcated. Both the form and content of statements articulated within the sanctum of the scientific community are seen as *radically* different from propositions advanced elsewhere" (1985, 31). The fruitfulness of such a reading will be demonstrated later in an analysis of the "cold fusion" controversy, in which just such a standard was deployed to delegitimate particular, highly controversial knowledge claims.

The Mertonian assumption of the scientific community's insularity is most clearly manifest in its proscriptions against public advocacy. Zuckerman insists that "going to the lay public for legitimation and recognition violates the norm of organized skepticism since it by-passes the primacy of qualified peer appraisal." She explicitly highlights the presumptive epistemic privilege of science when she adds, "Scientists remain uneasy about laymen being even implicitly regarded as though they were qualified to pass judgment on the substance of scientific work" (1977, 122). This is not to say that persons without particular sets of skills *should* evaluate products of those skills. Rather, it is to challenge the implicit presumption that unbridgeable barriers exist between scientific and social ways of knowing. The implications of enacting such a presumption will be illustrated in chapter 5, in an analysis of the response of the traditional scientific community to so-called scientific creationism.

In general terms, then, the Mertonian "paradigm" takes as given (indeed, necessary) the privilege granted to contemporary scientific "ways of knowing." Collins and Restivo argue that it "uncritically affirms modern science as the standard for objective inquiry. . . . It tends to function as an ideology for science as it is, with all its social trappings" (1983, 189). Merton himself suggests that the fundamental motivation of communal self-reflection was to craft more effective responses to perceived attacks from the "outside," saying, "An institution under attack must reexamine its foundations, restate its objectives, seek out its rationale" (1973, 115). This, of course, suggests the intrinsically rhetorical nature of the paradigm, its character as the functional use of discourse. It becomes problematic only when that character is effaced, that is, only when it is cast in a Mertonian rhetoric of self-denial. Viewed in this light, the Mertonian program is uncomfortably consistent with orthodox philosophies of science which insist that science is, and must be, a privileged source of knowledge.

Recall that Merton and his colleagues have argued that the normative structure of science is perhaps the most fundamental and active constraint on scientific practice. As such, they certainly imply that practicing scientists abide (consciously or otherwise) by such normative strictures. It has been argued often and persuasively, however, that just the opposite might be the case (Barnes and Dolby 1970; Mitroff 1974; Mulkay 1975, 1979).

Mitroff's study (1974) of the Apollo astronauts gives this argument perhaps its most compelling expression. He suggests that contrary to assumptions of the technical and moral norms identified by Merton and others, the Apollo scientists were personally biased, subjective, dogmatic, and secretive in their research activities. In short, they were acting as humans and did not consider those behaviors as at all non-scientific. The most immediate implication of Mitroff's argument is clear: that scientists, far from being bound in any absolute way to the posited normative structure, self-consciously act in ways contrary to that structure in certain circumstances. If such is the case, then the normative structure of science, as articulated by Merton, would appear either to be skating on very thin empirical ice or to be actually a contextually variable phenomenon, or both.

Zuckerman dismissed this critique, arguing that the issue of scientists' actual behaviors is not telling because "norms are, of course, not behavior. . . . they are standards taken into account in behavior and standards by which actual behavior is judged" (1977, 123). This seems to me a less than satisfactory response. Especially in the context of Zuckerman's stated concern with the internalization of particular norms as a means for the suppression of deviance in science, the apparent disjunction between norms and their practical manifestations represents a troubling empirical and theoretical conundrum.

Her second response to the issue is more substantial, but equally troubling. She insists that while the Apollo scientists might well have engaged in normatively unsanctioned behavior, they did so only in the domain of individual practice. She maintains that "of course, the norms of science refer to public science, not to the private phases which scientific investigators work through to arrive at results they are prepared to submit to their peers as justified claims to a scientific contribution" (Zuckerman 1977, 123–24). In sum, the initial phases of experimental practice either exert a different set of normative constraints on scientists or allow them to maneuver without the constraints of any articulable normative structure.

This too seems a less than compelling response. First, it trades indirectly on Reichenbach's untenable "two contexts" distinction. Demarcating the contexts of discovery (experimental/technical practice) from justification (peer evaluation, publication, etc.), Zuckerman implies that "nonmaterial" considerations such as normative constraints have no influence on the experimentally produced "substance" of scientific knowledge. Such a position reinforces portrayals of science as an insular and insulated sort of activity, that insularity being taken as the primary point of distinction between it and other sorts of activities. It could be argued just as plausibly that the technical or professional interests implicated in experimental stages of Apollo practice differed from those in other stages and called for adaptive rhetorical practices. Putting the argument in these terms requires jettisoning neither the constructed uniqueness of science nor its essential continuities with other social practices.

A second point at which the relationship of the normative structure and practices of science has come under critical fire involves the relative linkage of norms and the reward system of science. Mulkay has argued that "there are no compelling reasons for regarding these norms as 'operating rules of science' . . . and conformity to most of the supposed norms and counter-norms of science is largely irrelevant to the institutional processes whereby professional rewards are distributed" (1975, 536).

The implication of Mulkay's position is that, without such linkages, the so-called norms of science function as, at best, interesting sociological observations, but certainly not as important constraints on scientific practice. In that way, they would reflect only the perceptual scheme of the Mertonian sociologist of science and would be utterly irrelevant to an understanding of science, unless recast in their contextual function as part of a "vocabulary of justification" (Mulkay 1991).

While I am sympathetic to the thrust of Mulkay's critique on this issue, he too errs in assuming that the connection must be born of an individual's internalizing the norms and acting upon them with appropriate institutional responses following. As public discursive currency, the norms could well function profoundly to structure scientific practice without evincing the sort of connections that Mulkay seems to want. The reformulation of the norms can be read quite comfortably as a rhetorical act, an attempt to adjust "ideas to people" with competing views of knowledge, practice, and technique. The cumulative effect is to "adjust people to ideas," as scientists are socialized to accept the norms as necessary for the fruitful accomplishment of their communal practices.

While sharing Merton's concern with the "scientific community," Ziman advances a quite different portrayal. Perhaps in part as a function of his own self-described "outsider" perspective on questions traditionally claimed by philosophers and sociologists, Ziman labels his work "amateur philosophy" (1968, xi), conceding that most of it was written in his "spare time" (1980, ix), away from his professional scientific activities. This strikes me as unnecessary (though not insincere) modesty.

As a practicing scientist, Ziman is initially concerned to distinguish science from other ways of knowing. He ponders, "What *is* science? How is it to be distinguished from other bodies of organized, rational discourse?" (1978, 2). Recognizing the history of confused and confusing speculation on the question, he suggests that "to try to answer the question . . . is almost as presumptuous as to try to state the meaning of life itself. . . . We must accept it, as the good lady of the fable is said to have agreed to accept the Universe" (1968, 1). More than simply presumptuous, this effort might be read as, at base, a well-intentioned fool's errand, suggesting that scientists go about their daily affairs "without pretending to a clear and certain notion of what Science really is. In practice, it does not seem to matter" (6).

As Ziman subsequently makes clear, however, it does matter and in important ways. Most notably, he maintains, traditional sociological accounts of the scientific community (e.g., Hagstrom 1965), have been hamstrung precisely by their failure to engage reflectively the definitional question. He argues that "only too often the element in the argument that gets the least analysis is the actual institution about which the whole discussion hinges—scientific activity itself" (Ziman 1968, 10).[4]

Beyond the professional sociological realm, however, Ziman highlights the social importance of the demarcation question. Embedding science in its social contexts, he insists that "the question whether a particular body of knowledge . . . is genuinely 'scientific' is severely practical; upon its answer may depend our lives, our futures, our sanity, or our happiness" (1978, 158). Given the epistemic authority of science in contemporary culture, invoking the labels of science or nonscience to characterize a given set of claims is an evaluative judgment of the highest order.[5]

Hence, rather than measuring a standard of demarcation by its relative fit with the philosopher's ahistorical guidelines, Ziman insists that we look to the various practices of science and its practitioners in their social contexts for answers, implying that there might well be multiple answers to, perhaps, multiple demarcation questions. For him, "the answer [to the demarcation question] is *empirical;* does not depend on

4. This is not to suggest that Ziman looks favorably on philosophical accounts of science. While they may have sustained a long tradition of debate on the demarcation question, Ziman characterized that tradition as "arid and repulsive." Most lamentable, he suggests, is philosophy's penchant for "jobbing backwards," telling scientists "how we ought to have derived our result if only we had known the answer before we began" (1968, 31).

5. While I do not want to develop the implications of this claim for a critical rhetoric of science here, it seems that Ziman is content to recognize the risks of such naming, rather than indict the more general power/knowledge hierarchy. On this, see Foucault 1977a, 1977b.

some logical criterion . . . but on the context in which it occurs" (1980, 31). I will detail Ziman's "empirical" answer below. For the moment, however, it bears mention that this is an important departure from philosophical prescription as well as from the a priori assumptions regarding the boundaries of the functionalist paradigm. While Ziman does not push his perspective in such a direction, it seems at least plausible that there are multiple answers to the demarcation question, each dependent, as Ziman puts it, "on the context in which it occurs."

This is not to say that Ziman articulates an open-ended conception of science. He has no qualms, it appears, about providing his own answer to the demarcation question. Put briefly, "Scientific knowledge must be public and *consensible* (to coin a necessary word)" (1968, 11). By *public,* Ziman appears to have in mind the requirement that science (as a body of propositions) is subject to the open scrutiny of the scientific community. By specifying that scientific knowledge must be consensible, Ziman excludes those propositions about which that community cannot, in principle, agree.

Understood as a social community (rather than as propositions), science is said to be characterized by the attitudes of those advancing the consensible claims in question: "It is a scientific statement if it is made within a scientific discipline in the furtherance of scientific knowledge" (Ziman 1980, 31). But how are we to recognize a scientific discipline, as opposed, say, to a political one? Ziman encourages the analyst to "study the attitude of its professional practitioners. A sure symptom of non-science is personal abuse and intolerance of the views of one scholar by another" (1968, 28). In this story, science is portrayed as the domain of serious but open-minded individuals concerned with achieving consensus on important questions and having little time or use for the less salutary aspects of human behavior.

At least for the purposes of constructing an empirical rhetoric of demarcation, this standard seems untenable (and, for that matter, unnecessary) on two important dimensions. First, it is at odds with a large body of literature which painstakingly details the more than occasional intolerance of scientists for competing positions. (See, e.g., Aronowitz 1988a, 1988b; Prelli 1989a; Taylor 1992.) Second, while perhaps less common than desired, it is not unheard of for scholars in the humanities, say, to avoid personal abuse and intolerance. In this sense, the most significant limitation of Ziman's "consensus/attitude" demarcation position is not that he is wrong to contend that there are important differences among social practices. The most significant limitation, as I see it, is that he posits an unproblematically identifiable (albeit consensual) scientific "icon" against which other practices may be measured and found epistemically wanting. While this epideictic emphasis on the behavior of scientists seems motivated more by wishful prescription than by his underlying empiricism, it needn't obscure the considerable heuristic value of Ziman's construal of demarcation as, at base, an empirical process that gets worked out in practice.

Like Merton's implicit focus on audience, Ziman's contextual view implicates the situational constraints impinging on rhetors and rhetorics of demarcation. Specifically, his analysis of the tension between innovation and coherence illustrates the practical contingencies confronting scientists and constraining their rhetorical practices. He notes, for example, that "the fact remains that [tolerance of dissent and critical evaluation] are the constitutive principles around which the modern scientific community now functions" (1978, 131), conceding simultaneously that "despite all its high ideals and good intentions, such a community must inevitably resist radically new ideas that upset its hard won position and throw into doubt all the earnest labours of its members" (42). This creates not contradiction but practical contingency—a condition that invites, even demands, rhetorical practice.

In what I take to be a significant departure from Mertonian orthodoxy, the range of contingencies includes the so-called public implications of scientific claims. Inasmuch as Ziman was trained as a physicist, it is not surprising that his intellectual allegiances lie with the scientific community. Nonetheless, he gives evidence of a concern for the public "image" of science, noting that "since the Republic of Science must never be closed, it is important that public justice be done to such [unconventional] ideas in open court. Nothing is more destructive to the credibility of science in the public eye than the appearance of scorning or suppressing anything that might, in any way, be considered a sincere, and possibly valuable, contribution" (1978, 143). Similarly, his discussion of pseudoscience concludes that "the manner in which such matters are internally handled may have considerable influence on the public image of science and, under pathological political conditions, may be highly relevant to our theme of reliable knowledge" (145).

These passages suggest that the pristine openness, tolerance, and so on that Ziman advocated as constitutive of the scientific community are not ontological judgments, but should be understood as contextually constitutive in the sense that they are deployed as symbolic resources in response to the contingencies brought on by the interpenetration of the so-called public and technical spheres of practice. Understood in this way, what we take to be science cannot be understood (in any coherent sense) apart from the complex web of social practices in which it is necessarily embedded.

In *Teaching and Learning about Science and Society* (1980), Ziman attempts to cash in this interpretive strategy by sketching out the various components of "academic science," including the academic scientist, the scientific community, and scientific knowledge itself.[6] Those elements are then empirically grounded in an overarching account of what he calls the "R and D" system, including aspects of "education and popularization," "technological innovation and development," "big science management," and "science policy and economics" (1980, 68–88). Subsequently, he embeds that model into its "social context," framed by "expertise," "ideology," "instrumentality" (e.g., economics and politics), and "responsibility" (89–107). Far from being a narrowly defined body of knowledge or activities, science is portrayed as "a way of doing intellectual business. . . . It is like the Stock Exchange" (1981, 32).

Like the stock market, hypersensitive to influences as disparate as a drought in an agricultural region and the prospect of a Democrat in the White House, science, in its broadest sense, is very much a product of the contextual activities, at both the macro- and microlevels, of its social constituents. It exists precisely at the intersections of those constituents. In Ziman's terms, what appears to be missing is how the linkages between those constituents and their multiple intersections are crafted. That too is an empirical question and one that is especially amenable to rhetorical analysis. The contextually contingent nature of the scientific community is underscored in Ziman's claim that "Science, with a capital S, is much too complex to be treated as a single 'thing.' . . . It is a large-scale human activity of many aspects and many intersections with other components of society in the intellectual, material, psychological, and political dimensions" (1981, 75).

The Sociology of Scientific Knowledge: Rhetorical Implications

Positioning Merton and Ziman as icons, I have so far traced a shift from the assumption of science's demarcated uniqueness to its enactment in critical and analytic practice. At both levels, however, recent intellectual developments have inspired attempts to understand science and scientific knowledge as social accomplishments. Primarily European in origin, these attempts have, among other things, taken seriously the proposition that science, as a set of knowledge claims

6. I am indebted to Andy Pickering for the metaphor of "cashing in," provided during a class session that he has no doubt long forgotten.

or as a set of social practices, is equally amenable to sociological analysis (Lynch 1992; Pickering 1988). While the following section will discuss the important distinctions among individual formulations of this position, the guiding commitment of the "strong" and "constructivist" programs is to the notion that the production of scientific knowledge cannot be understood meaningfully without close attention to the social contexts of and constraints on that production. Bohme and Stehr have argued that "production of knowledge, in the sense of an immediate productive force, becomes a social resource with functions comparable to those of labor in the productive process" (1986, 12). More broadly, as Woolgar put it, "It is not that science has its 'social aspects,' thus implying that a residual [hard core] kernel of science proceeds untainted by extraneous non-scientific [i.e., social] factors, but that science is itself constitutively social" (1988b, 13).

The implications of such a position for the current project are considerable. Most notably, there is a clear indication that science is not vested with some sort of ahistorical, transcendent "essence." It is rather the product of particular social, historical, but above all practical configurations of persons and equipment. Those configurations have a human, hence rhetorical, rather than natural ancestry.

The story of the sociology of scientific knowledge is a complex one, complete with alliances, counteralliances, and "schools" of various stripes (Lynch 1992, 215–24; Pickering 1992). There are different strains within the "constructivist" program (e.g., Mulkay and Gilbert 1982a, 1982b; Knorr-Cetina 1981a, 1981b; Latour and Woolgar 1979), each with its own approaches to the sociological study of science and scientific knowledge. While particular research strategies differ—for example, ethnographic study versus discourse analysis—the constructivist accounts hold that once one studies the activities and discourses in research settings over a period of time, the socially constructed natures of what comes to count as science, knowledge, and scientific knowledge are illustrated. Consequently, rather than scientific knowledge claims being defined relative to practice-independent criteria, scientific knowledge is seen as "manufactured" (Knorr-Cetina 1981a) in ways similar to other goods. The constructivist program (hereinafter CP) is grounded in scientists' in situ practices, with a particular strain concerned with discursive practices (e.g., Mulkay and Gilbert 1982a; Yearley 1985). The following sections offer more specific discussions of these approaches and their mutual relevance to the conceptual and methodological background for the rhetorical study of demarcation.

Barnes, Bloor, and the Relativist Program

The organizing problematic for the relativism of Barnes and Bloor is this: If observational considerations alone cannot establish the cognitive or epistemic significance of a given belief system, then what else does? Kuhn's (1962, 1977) by now commonplace argument that the production and legitimation of scientific knowledge are conceptually inseparable from the institutionalized norms and explanatory paradigms accepted by members of a given community provides the starting place for their theorizing, though they do not interpret Kuhn's "nature and logic" as socially invariant and universally binding variables. Barnes, for example, argued that "once beliefs are conceded not to derive from the constraints of reality no further *a priori* argument can be made against their sociological investigation. And the problem of validity of the beliefs in question is beside the point in this regard" (1974, 12).

The relativist program, then, extends dramatically the implications of the "underdetermination thesis." Maintaining that there is little justification for sustaining the presumptive superiority of scientific knowledge over other forms, Barnes noted, "The culture of natural science cannot be distinctive because of its rationality in a universal, rather than in a conventional, sense" (1974, 41). The key recognition here is that knowledge claims are interpretable as culturally generated and sustained artifacts. What distinguishes knowledge claims from material artifacts is the propositional and discursive nature of the former.[7]

This open-endedness highlights the close relationship between the demarcation question and questions of the legitimation of knowledge claims. Bloor (1976), for example, maintains straightforwardly that "knowledge" (of whatever variety) is simply whatever a given human collectivity endorses as constituting knowledge. He argued that "the knowledge of a society designates not so much the sensory experiences of its individual members. . . . It is rather, their collective vision or visions of Reality. Thus knowledge . . . as it is represented in science . . . is better equated with Culture than with Experience" (12). Read in this way, it is futile to consider the epistemic privilege granted a particular variety of knowledge claims apart from the constitutive (and communally validated) understandings of the nature of that knowledge. This is fundamentally a question of communally produced and sustained demarcation practices.

A second important implication of Barnes's and Bloor's relativism is that it calls into question the processes wherein scientists themselves

7. We shouldn't assume, however, that the discourses of science are somehow simple reflections of material artifacts.

negotiate the communal standards they invoke to define and evaluate their own practices. Bloor, for example, noted the epistemic tar pits into which philosophers have fallen in an attempt to find some principled justification for the formal criterion of "simplicity" and pondered its implications for scientific practice: "What gives simplicity its credibility for this but not that group? . . . It is more plausible to see simplicity as an 'after-the-fact' justification for opinions that have their real basis elsewhere" (1981, 201). If we take his claim seriously, we can extrapolate that the presumably constitutive standards applied by scientists are themselves the products of rhetorical negotiation. As such, the demarcation of science is not a task confined to philosophical ruminations. Rather, it is a rhetorical accomplishment of those rhetors we take to be scientists, and the disparate social actors with whom they interact in the conduct of ongoing scientific practice.

While recognizing that social practices do differ, Bloor insists that they need not be seen as doing so in evaluatively different ways. He writes that "there is nothing special about science that resides in the biology of scientists. . . . There is not even anything special about their using pieces of apparatus or their taking measurements and samples. It is their goals and the interpretations that they put on their interactions with the world that matter" (1981, 199).

In this sense, science, or that series of social practices which has been given the communal designation thereof, is not intrinsically distinct from or superior to the broader culture. Quite the contrary. From this perspective, science cannot be understood sensibly apart from that culture. As Barnes noted, "One cannot identify distinct kinds of knowledge and culture, but only distinct modes of use of the culture we share. What we take as our current scientific knowledge is that of our overall culture which is currently licensed for technical use" (1982, 37–38).

This interpretive stance suggests that demarcation practice need not involve identification of the "timeless" markers of science, but rather involves the historical negotiation of the role that science is to play within a particular cultural configuration. Accordingly, we need not and should not unreflectively assume that any such role will remain stable throughout its interactions with those cultural configurations (Fuller 1992; Ben-David 1981, 1984). Just as our understanding of our broader systems of cultural practices is multifarious, so too is our understanding of science and its relations to those practices. Hence, my reading of this literature takes as a virtue the occasionally inchoate, always open-ended character of science.[8]

8. In the true spirit of the postmodern, the postmodern condition (Lyotard 1984) has been described and evaluated by a number of theorists. In some ways, I draw only on the

It is important to avoid a common point of confusion on this issue. While Barnes, Bloor, and others do self-consciously articulate a relativist position, it should not be taken as a simpleminded relativism. Rationality is not construed as irrelevant; it is the nature of that rationality that is at question in this discussion. Barnes contends that "in asserting the insufficiency of 'reason' in science, [I] will in no way imply that scientists are unreasonable men; rather it will oppose an intolerably individualistic conception of cognition. The argument will not be that something the opposite of what is rational guides . . . science, but that the entire framework wherein the reasonable and the social stand in opposition must be discarded" (1983, 22). At the very least, as Woolgar puts it, "the notion of 'social' [similarly 'cultural,' 'psychological,' and so on] has to be substantially modified" (1988b, 13). How broadly we cast the "social" is itself a discursive accomplishment.

This is not to say that all questions of scientific controversy are necessarily influenced by such national or dynastic issues. As Bloor explains, "The question of the scope or kind of the social factors at work . . . is entirely contingent. The important point is that where broad social factors are not involved, narrow ones take over" (1981, 203). The "narrow" social factors to which Bloor refers can be understood as deriving from the interests, traditions, or routines of the scientific community. The rationality of science (and other forms of culture) is contextually and historically situated and is mediated in and through the discursive practices of scientists.

The Strong Program

The foregoing discussion of Barnesian relativism serves as conceptual background for the more specific commitments of the "strong program" (hereafter SP). Here then, I will sketch in very broad strokes the major tenets of the SP and their implications for the current study of demarcation rhetoric.

Barnes's and Bloor's relativist construction of scientific knowledge as a cultural artifact which is defined and sustained by participants' collective practices serves as the background for the four theses associated with the SP.

First, sociological accounts should be causal, "that is, concerned with the conditions which bring about the belief or states of knowledge." While recognizing the pervasive sociality of human practice, Bloor

broadest assumptions common to most articulations of postmodernism. For broad discussions of postmodernism, consult, for example, Lash 1990.

notes that "naturally there will be causes other than social ones which cooperate in bringing about belief" (1976, 3). The SP does not exclude empirical or institutional factors; these factors are simply made contingent on social processes of interpretation, evaluation, and so on.

The second thesis entails that sociological accounts must be impartial with respect to truth and falsity, rationality or irrationality, success or failure. In short, those beliefs thought true, for example, that the earth resembles an oblate spheroid, cannot be treated differently from beliefs such as those of flat earthers, that are subject to widespread criticism, even ridicule. Of primary significance here is that the SP abandons a priori classifications of knowledge claims and their relative amenability to sociological explanation. Significantly, the truth or falsity of said claims cannot be advanced to justify a given scientific practice.

The third thesis is that an adequate sociology of scientific knowledge must be "symmetrical in its style of explanation. The same types of cause would explain, say, true and false beliefs" (Bloor 1976, 4–5). Related to the impartiality thesis, it actively militates against the traditional assumption that "true" scientific knowledge earns its status through a direct and rather unproblematic relationship with objective reality while false belief is a product of "social contamination." The SP argues that social influences (of whatever variety) actively constrain knowledge claims which are ultimately granted or denied the imprimatur of communal acceptance.

The final thesis of the SP is that its propositions must be reflexive. In essence, that which is good for the scientific goose is equally good for the sociological (and rhetorical) ganders. Bloor explains that "it is an obvious requirement of principle because otherwise sociology would be a standing refutation of itself" (1976, 4). While some have argued that the SP is such a standing refutation, this thesis is itself an indication of the SP's recognition of the ultimate indeterminacy of knowledge claims.[9]

The SP, on the other hand, avoids the interpretive lacunae of traditional accounts of science as bound by nonconventional criteria. Barnes concedes that the SP "is sceptical because it suggests that no arguments will ever be available which could establish a particular epistemology or ontology as ultimately correct" (1983, 154). In this sense, the SP insists that such distinctions are, at base, a matter of (nontrivial) conven-

9. In more recent work, this emphasis on reflexivity has been manifest in the increasing prominence of "alternative discursive forms" (Ashmore 1989). These forms exploit the ultimate indeterminacy of texts (whether of science or social studies of science) in imaginative ways. It is not clear, however, that they resolve the problem of reflexivity, or that it should be "resolved" to start with. Whatever scholarly convention one chooses to follow (or flout), we must remember that we might be wrong.

tion. Scientific evaluation, then, is a matter of "judgment at the individual level, of agreement at the level of the community; it is open-ended and revisable" (Barnes 1983, 30). The fundamental constitutive role of demarcation negotiation is apparent in Barnes's claim that "it is not that knowledge is a system of conventions which determines how we think and act. On the contrary, it is our decisions and judgments which determine what counts as conventional" (30). The demarcation of science, then, is necessarily and intrinsically a rhetorical matter.

By implication, the judgments which emerge from local disputes are central to the construction of the frameworks enabling us to understand the nature and meaning of science. In a related vein, Harry Collins (1981a, 1981b, 1985) insists that controversy is the context in which the analyst might most clearly see how competing claims and assumptions are stabilized and destabilized over time. We come to understand demarcation as a rhetorical matter when "the constitutive picture of science is put under some strain for the actors. This is what happens in scientific controversies. The controversy is an *'autogarfinkel'* for the world of scientific knowledge. Underlying processes of social construction are made manifest to actors" (Collins 1981a, 14–15). Even when controversies appear comfortably in the domain of science, that appearance is a matter of boundary reification.

It is important that such controversies need not be limited to rather clear-cut cases of science versus "pseudoscience." If Collins' claim is accurate, any sort of scientific controversy ought to afford a wealth of evidence of the rhetorical practice of scientific definition and legitimation, that is, evidence of demarcation practices (Bohme 1979). The two case studies to be featured in chapters 5 and 6 will attempt to cash in the interpretive strategy of examining scientific controversies as demarcation episodes in different contexts.

At base, the SP seeks to open up scientific practice and knowledge production to sociological investigation. For the purposes of the perspective I want to develop, it suggests a systematic account of the social production and legitimation of knowledge claims within the community that we have come to take as "science."

While the SP is a valuable intellectual resource for enriching our understandings of science and scientific knowledge, it is certainly not without its critics (Gieryn 1982; Hollis 1982; Jarvie 1983; Lukes 1982). For the most part, however, those criticisms are not relevant to this study, and I will bracket them out. One criticism, however, is of particular consequence for my narrative and warrants attention.

As I noted above, the SP purports to be a "scientific approach" to the sociology of scientific knowledge. In many ways, this is a troubling claim. The proponents of this program extend defensible relativist commitments to the content of scientific knowledge claims, to scientific practices, and even to questions of reflexivity and objective reality. They seem to pull up short, however, when it comes to the extension of those commitments to the nature of science itself. At no point does the SP offer a principled account of what it takes the nature of science to be; it is left to us to assume that the SP somehow fits the model. Larry Laudan argued that "one [should] postpone any decision about whether the strong program was scientific until after one had carefully studied—sociologically or otherwise—what features the knowledge systems we call 'scientific' exhibit" (1981, 183).

While I do not accept the bulk of Laudan's criticisms of the SP, this seems a telling indictment. I disagree, though, on the prescribed remedy, one that appears considerably worse than the malady itself. Rather than, like Laudan, nearly rejecting wholesale the SP, I am inclined to extend the SP's insights to its own purported object of investigation. If we argue that the nature and meaning of science, as social practice, is rhetorically defined and sustained in social and rhetorical practice, then that would seem to be a reasonable approach. I suggest that we not restrict the areas in which its relativism obtains, but rather extend them one important conceptual step further.

It should be clear from the above discussion that the SP, as articulated by Barnes and Bloor, is cast at a fairly broad conceptual level. Indeed, a characteristic of much of this work is its programmatic and theoretical nature. The interpretive strategy is only rarely cashed in in observational or textual analysis (Collins and Restivo 1983). It is precisely this limitation to which the emergent "constructivist Program" in the sociology of scientific knowledge responds.

From Sociological Theory to Interpretive Practice

While enacted in a number of quite different ways, the CP is concerned fundamentally with precisely those sorts of localized investigations. Knorr-Cetina (1981a, 1981b), while generally sympathetic with the SP, has suggested that the most fruitful approach for the sociology of scientific knowledge is to assume the sociality of science and to work toward documentation of how that sociality is manifested in scientists' reasoning and experimental practice.

A host of studies can be comfortably grouped under the rubric of the CP. To address each of these formulations and their subtle nuances specifically would be a herculean task, and one better left to a sociologist with quixotic aspirations. Accordingly, I want to sketch briefly the broadest common commitments of the CP and then demonstrate their relevance and importance for the current study of demarcation by close attention to exemplars of the tradition (Knorr-Cetina 1981a; Latour 1983, 1987; Latour and Woolgar 1979). In so doing, I want to advance the argument that the CP's focus on the empirical practices of science provides a highly practical (and "textual") dimension to the epistemological commitments of the SP, gesturing toward an important conceptual and methodological direction for a rhetorical study of demarcation. That direction leads reasonably to a consideration of the variable, indeterminate, and strategic dimensions of the deployment of particular demarcation criteria.

In contrast with traditional accounts of science associated with traditional philosophies and sociologies of science, the CP utterly rejects the classical dichotomy between the production and legitimation of scientific knowledge. In short, it is precisely within the deployment of local practices, discursive and otherwise, that the grounds for epistemic warrant are situated. As Tibbetts puts it, "On this account scientific knowledge is a contingent achievement, manufactured in variable interpretive contexts" (1986, 167). This general commitment extends beyond ontology to broader questions of legitimacy and epistemic warrant. Within these, the CP's relevance to the demarcation issue is highlighted through the suggestion that the most fundamental basis of claim-making and judgment, their relative "scientificness," is itself a social construction.

The CP enacts a research strategy centered largely in fine-grained textual and field studies of scientists' interpretive practices in local laboratory settings. In effect, that which is taken to be cognitively and socially important about science and scientific knowledge production is to be found at the sites of scientific production and evaluation. As Pickering put it, "Scientific knowledge has to be seen . . . as knowledge relative to a particular culture, with this relativity specified through a sociological concept of interest" (1992, 5).[10]

10. The interests that Pickering identifies as critical to the production and evaluation of scientific knowledge may, for example, be predominantly political, communal, or economic. Whatever the particular motivation for a particular actor (rhetor), it is crucial to recognize here that the purveyors of scientific rhetoric are construed as active agents, rather than passive viewers of the mirror of nature. Pickering puts it nicely: "Knowledge is for use, not simply for contemplation, and actors have their own interests that instruments can serve well or ill" (1992, 4). What is missing here, it seems, is an account of how those interests are advanced, defended, negotiated, evaluated, and closed down—a matter of rhetorical analysis.

At first glance, it seems paradoxical that such an observational commitment is grounded in a thoroughgoing relativism. Indeed, Woolgar (1986) has observed an irony in the CP's advocacy of a relativist epistemology for reconstructing scientists' reality accounts and an epistemological realism regarding its own observational accounts. On this issue, it is helpful to recognize at the onset that the tension between constructivists and their realist critics is not, strictly speaking, concerned with ontology. Constructivists needn't engage questions of existence, as they are concerned instead with the elasticity of scientists' accounts of reality. As such, the CP's commitment to the premise that empirical knowledge and the structures of its production are social achievements entails at least the possibility that such knowledge is contingent on and inseparable from the contextualized interpretive conditions of its production. In short, the focus is a discursive one, on the language practices within a given community of scientific practitioners.

Latour and Woolgar's study, *Laboratory Life: The Construction of Scientific Facts* (1979), is perhaps the most extensive and representative example of the CP's research commitments and strategies. This two-year case study focused on scientific research and journal accounts of the neuroendocrine TRF(H), or Thyrotropin Releasing Hormone, in a biochemical laboratory. The selection of such a research site was not merely incidental. They explain: "We have chosen to study the historical genesis of what is now a particularly solid fact . . . [one which] would hardly seem amenable to sociological analysis. If the process of social construction can be demonstrated for a 'fact' of such apparent solidity, this would provide a telling argument" for the utility of constructivist research (106).

We should clearly understand the moral of the story they tell. Latour and Woolgar were not concerned to enter the debate regarding the true nature of TRF(H). They were concerned rather with the processes by which its meaning was socially (re)constructed over time and the experimental significance attached to it as a result. By bracketing preliminary ontological assumptions, they hoped to present an argument that "facts" about the substance were comprehensible in social and interpretive terms. The stories of creationism and cold fusion will be similarly comprehensible in rhetorical terms.

Of course, Latour and Woolgar would not bill themselves as rhetorical critics; the means to their ends were anthropological. Latour, who actually conducted the on-site investigations, wrote that "the focus of our study is the routine work carried out in one particular laboratory. The majority of material which informs our discussion was gathered from *in situ* monitoring of scientists' activity in one setting" (Latour and Woolgar 1979, 27). As such, their work embodies the constructivist commitment to the minutiae of scientific practice as the central influences on scientific progress, rather than the epochal revolutions that typically receive sociological attention. Latour and Woolgar's strategy was, then, to "shed light on the nature of the 'soft underbelly' of science" (27), that is, on the everyday productive and evaluative practices of scientists.

While not even implicitly a rhetorical account, this is heuristic in its implication that broad esoteric ruminations on the appropriate nature and practices of science are less relevant to its understanding than are what scientists actually do and what they actually say about what they do or have done. This is a conclusion quite congenial to a rhetorical perspective on demarcation. To suggest that the demarcation of science is a rhetorical matter worked out, in part, by scientists in the course of everyday practice is equally to suggest that it is from scientists' social accounts of that practice that evidence must be accumulated. While the relative importance of participant observation versus analysis of verbal accounts remains unestablished in the sociological literature, the general methodological commitments of the CP provide compelling support for this project.

A key implication of this research strategy is that "laboratory life" is implicitly conceptualized as a culture. That culture is defined by the practical and symbolic activities of its members. Construed in this way, a cultural perspective on science embodies a means to the end of decreasing the unreflective privilege granted "scientific" practice simply because it is "scientific," in favor of making that privilege the outcome of more self-conscious and empirically grounded understanding. Such a perspective on scientific culture is "intended to dissolve rather than reaffirm the exoticism with which science is sometimes associated. . . . Scientists in our laboratory constitute a tribe in danger of being misunderstood if accorded the high status with which its outputs are sometimes greeted by the outside world" (Latour and Woolgar 1979, 29-30). To confuse the messenger with the message in this way distorts the nature and functions of both.

A second implication stems from the way in which "culture" is actually operationalized in Latour and Woolgar's anthropology of the laboratory. They define *culture* as "the set of arguments and beliefs to which there is a constant appeal in daily life" (1979, 55). The relevance of this understanding of culture to a rhetorical view of demarcation is clear. The presumptive constraints on scientific experimental practice are seen as intrinsically rhetorical. To the extent that argumentative resources provide the justifications for the selection of the means of knowledge production, these beliefs are directly and necessarily linked to the content of scientific knowledge claims. While I focus on the most fundamental class of that "set of arguments," the class which defines the constitutive nature of the community for its members, Latour and Woolgar's account would seem congenial to a broader commitment to the symbolicity of scientific practice. Given the disciplinary concerns they advance, it is perhaps not surprising that they do not capitalize fully on the provocative nature of this position. Concerned more with the construction and justification of claims, they do not pry open the related "black box" of culture-constitutive presumptions.

Despite this thematic distinction, it is important also to recognize that the story Latour and Woolgar tell is one full of political intrigue. Scientific practice, encompassing the move from material action to the evaluation of resultant knowledge claims, is situated in what they label the "agonistic process or field." This refers to the sum total of members' political negotiations, overstatements, and controversy over modality assignment. Latour and Woolgar, then, collapse, for all meaningful purposes, the social and political, crafting a space suitable for *krisis*, critical praxis. As they put it, "Once it is realized that scientists are oriented toward the agonistic field, there is little to be gained by maintaining the distinction between the politics of science and its truth" (1979, 237).

In some ways, the above can be taken as little more than a less direct way of saying that "science uses rhetoric." It is significant, however, to consider that sociologists, particularly those who fancy themselves scientists, are not often given to such proclamations. Indeed, one gets the sense that Latour and Woolgar (and many sociologists, such as Cozzens and Gieryn [1990, 5]) find this whole rhetoric business fashionably subversive, rather than a move toward reconstruction. Whatever their personal feelings, it is important to consider that their analysis provides compelling justification for the "leveling" of the discursive practices of science to a plane at which they can be meaningfully understood along with other sets of social practice. As they put it, "the negotiations as to what counts as a proof or what constitutes a good assay are no more or less orderly than any argument between lawyers or politicians" (Latour and Woolgar 1979, 237).

Taking Latour and Woolgar as sufficiently representative examples of a widely disparate research tradition, it is clear that the CP articulates a broad and interpretively democratic perspective on the nature and functions of science. As a whole, it argues persuasively for rejecting narrowly demarcated conceptions of science in favor of multiple demarcations constructed and sustained for multiple reasons.

Knorr-Cetina urges abandonment of the traditional unitary scientific community. In her view, science as a social practice is cast as "trans-scientific fields," referring to "networks of symbolic relationships which in principle go beyond the boundaries of the scientific community" (1981a, 82). Woolgar casts his analysis at an even broader level, maintaining that "there is no such thing as 'science' or 'the scientific method' except that which is variously and multiply attributed to the particular textual purposes for which this is an issue" (1988b, 21). Latour's (1983) case study of Pasteur and the nineteenth-century debate over anthrax illustrates this general commitment. He argues that through experimental findings, on-site demonstrations, and public debate, Pasteur variably reconstructed the concept of the microbe, and even the concept of laboratory science itself. Of especial relevance to the current study is the way in which, through a series of material practices, a statement about one object (microorganisms) was socially reconstructed into and accepted as a statement about another type of object (anthrax disease in macroorganisms, i.e., cows). Latour implies that the traditional distinction between what is internal and external to the laboratory is destabilized in such cases. More to the point is that the similar boundary between the cognitive and the social was also deconstructed. Through changes from the laboratory to the farm to the laboratory, "no one can say where the laboratory is and where the society is" (Latour 1983, 154). Given the agonistic process, it becomes clear that no principled and objective demarcation can be drawn between sites of scientific and social practice. This is not to say, however, that demarcation cannot be accomplished in rhetorical practices. What all this points toward is a view of science in which demarcation is a product of the interfaces between scientific practitioners and other social actors.

The textual process wherein these interfaces are played out, it seems to me, is a matter of rhetoric. Consider, for example, Latour's (1987) extensive discussion of enrolling "allies" as a means of gaining the requisite resources for scientific practice. In this case, the practice of science belies the traditional image of the solitary scientist puttering asocially away in a laboratory that provides grist for Hollywood and science textbooks. Latour aptly described the interconnected nature of science, saying, "When scientists appear to be fully independent, surrounded only by colleagues . . . it means they are fully dependent. . . . when they are really independent, they do not get the resources with which to equip a laboratory [or] to earn a living" (1987, 158). The apparatuses that form the basis for so much of our advanced research are made possible only through the rhetorical negotiation of a relatively stable supply of funds, whether from corporate interests or the federal government via the Department of Energy, the National Science Foundation, the Department of Defense, and the like. As a consequence, the phenomenon (or phenomena) that we call science transcends the balkanization we too quickly accept in modern culture, maintaining that our most basic analytic categories—science, public, and government—are themselves in dire need of revision.

To say that there is no principled difference between sets of social practice is not to say that there are no functional differences. At the very least, scientific practices are represented publically as somehow distinct from other practices. It seems defensible to concede that few rhetorical critics or politicians don lab coats and grind rat brains or install TEA laser components (H. Collins 1985). Indeed, my goal in this study, to illustrate how contextual boundaries between science and other practices are constructed, implies some sort of difference.

The major cause of dispute here is that the CP, because of its own disciplinary concerns, has been imprecise in its discussions of the essential continuity of scientific and social practices, as opposed to the "social use" of those practices. It is not necessary to suggest that there is absolutely no functional or evaluative difference between science and other practices in a world in which the National Science Foundation rarely funds avant-garde performance art or rhetorical criticism. Paradoxically, the CP seems to have neglected the social difference between science and other practices. It is precisely that difference which lies at the root of its (perhaps unreflective) epistemic privilege. While it seems wise to reject the traditional Mertonian conception of the "scientific community" (Dolby 1982), it is the worst sort of critical myopia to insist that socially constructed differences do not exist and do not influence our social lives in fundamental ways. If nothing else, the status of "science" as sociocultural symbol merits recognition and the critical interrogation that a rhetorical account demands and provides.

In this section I have read the "constructivist program" in the sociology of science through a rhetorical lens, in search of a rhetorical account of demarcation. I have argued that the CP's focus on close investigation of specific research contexts provides compelling evidence for close textual studies of science and scientific practice. Additionally, I have argued that the variable cultural presumptions and identities of science emerge from historically conditioned processes of argumentation.

An implicit theme throughout this discussion has been the utility of scientists' discourse as an object of investigation. Knorr-Cetina, for example, suggested that scientists' discourse, both written and oral, has important effects "in the concrete negotiations of the laboratory . . . and leads to the continuous reconstruction of knowledge" (1981a, 14). Other work consistent with the CP (e.g., Lynch 1985a, 1985b, 1992) has speculated on discursive processes in science as well.

I want to engage this sociological interest in discourse analysis in order to highlight what I take as its unique theoretical insights, as well as its limitations. Those limitations, many born of quite reasonable disciplinary differences, bring into sharp focus the interpretive and political value of making the turn to rhetoric.

Discourse and the Social Construction of Science

Given the complex nature of scientific practice, it is not surprising that the multitude of methods for its investigation yield complex insights while raising equally complex questions. The discourse analytic approach to science studies has developed primarily as a methodological alternative to formalist studies of the scientific community, bibliometric surveys, and ethnographic observation. As Mulkay argues, "The analysis of discourse is being presented as an alternative to the more traditional concern with describing action and belief" (1981, 163).

At base, the point of contention is whether more or less literal descriptions of actions or practices provide sufficient or defensible evidence on the nature of science. Discourse analysts suggest that such descriptions cannot begin to justify the insights that their advocates contend. Mulkay argued that "the traditional objective of describing and explaining what 'really happened' has been abandoned and replaced with an attempt to describe the recurrent forms of discourse whereby participants construct their versions of social action. One focusses . . . on the methods scientists themselves use to account for and make sense of their own and others' actions" (1981, 170). Understood in this context, the analysis of scientific discourse rather self-consciously embraces a view of (the meaning of) science as a social construction whose parameters and constraints emerge from the discursive practices of scientists—practices that, in short, are not means to an end, but critical ends in themselves.

The most sophisticated exemplar of this work, Gilbert and Mulkay's *Opening Pandora's Box* (1984), does pay homage to traditional discourse analytic scholarship as a sort of methodological soulmate. Gilbert and Mulkay contend that their work bears important similarities to the sociolinguistic tradition, in that "it attempts to provide a systematic description of discourse employed by particular social actors in special settings" (16). Those special settings have included research notes (Garfinkel, Lynch, and Livingston 1981), research articles (Myers 1985), and even science lectures and textbooks (Yearley 1985).

As operationalized in the sociology of scientific knowledge, *discourse* is used as a "convenient way of referring to all forms of verbalization: to all kinds of talk and verbal data" (Mulkay 1981, 169-70). A major implication of this conceptualization is that no particular forms of discourse are considered analytically prior or superior to another. As Gilbert and Mulkay have suggested, "The informal talk whereby actions and beliefs are constituted at the laboratory bench is not regarded as having primacy over any subsequent interpretation around a coffeetable, at a conference, in a research paper, or at an interview" (1984, 14). Accordingly, discourse analysts allow themselves a broad and diverse sample of discourse on which to apply their particular methods of analysis.

By implication, the discourse analytic program can be understood as a straightforward attempt to account for the apparent complexities and diversity of constructed scientific realities (Fuhrman and Oehler 1986). This represents a significant departure from previous research, which was organized around something of an "apologetic" theme, the impetus of which appeared to be the demonstration of the literal veridicality of scientists' discussions of the natural world: "Sociologists' attempts to tell *the* story of a particular setting or to formulate *the* way in which social life operates are fundamentally unsatisfactory" (Gilbert and Mulkay 1984, 2; see Lynch 1982, 1985a, 1985b). It appears that these attempts fail because they "imply unjustifiably that the analyst can reconcile his version of events with all the multiple and divergent versions generated by the actors themselves" (Gilbert and Mulkay 1984, 2).

It is important to note that discourse analysts do not claim to have themselves discovered *the* definitive account of social practice that had eluded all other attempts. It is true that various authors in this tradition may make more or less temperate assertions regarding the representational accuracy of their claims (Lynch 1985a; Potter 1984; Yearley 1985). The methodological payoff for this approach, however, is to be found in the insights generated in and from the multitextual nature of scientific practice. This rhetorical sensitivity is a crucial development in the sociology of scientific knowledge, insofar as it insists on a multitextured approach to the investigation of a multitextured phenomenon, science as social (discursive) practice.

The key theoretical characteristic of scientific discourse is its variability. Rather than attempting to discover or construct a literal account of scientific practice, discourse analytic approaches investigate alternative constructions of reality which emerge from alternative discourses, presuming that the very possibility of one definitive construction of science is chimerical. Within its variable discursive contexts, the constitutive natures of science and its knowledge claims are shaped.

This brief description of the discourse analytic approach in the sociology of scientific knowledge indicates its relative improvement over previous research strategies and its relevance to the current study. Initially, discourse analysis eliminates the thorny epistemological and theoretical conundra which emerge from a view of science as providing the single appropriate view of reality. Interpretive variability, even indeterminacy, raises its own set of difficulties. The discourse analytic view, however, treats such variability not as a source of error, but rather as a reflection of the theoretical richness afforded by a discursively constructive understanding of science. Mulkay insists that "all the detailed inconsistencies between accounts which occur in all qualitative analyses cease to be specially troublesome . . . once one stops trying to get through to what really happened" (1981, 170).

Rather than attempting to "explain away" the richness of discursive variability, discourse analysts insist that meaningful attention must be given to the contextual constraints which lead to the manufacture of that variability. Indeed, the "difficulty" represented by such variability is experienced only by analysts seeking to offer definitive accounts. For scientific practitioners (and discourse analysts), such variability is not a difficulty at all, but a mundane constraint on the multiple practices of science. As Gilbert and Mulkay emphasize, "Interpretive uncertainties . . . do not pose any great difficulty for participants, who have at their disposal a range of flexible techniques which enable them to make sense of whatever is going on in a way that is adequate for most practical purposes" (1984, 13).[11] The contextual production and deployment of demarcation criteria represent the "practical purpose" interrogated here.

This contention is given even more support when one considers the role that discourse analysts attribute to the object of their studies. As conceived by Gilbert and Mulkay, discourse analysis "is a study of aspects of scientific culture which documents some of the methods used by scientists as they continually construct and re-construct their social world" (1984, 17). As such, science is construed as a culture which can only be understood through its constitutive practices. The meanings of those practices are constructed in and through the discourse of scientists.

Such an understanding is strikingly similar to my call for a constructively rhetorical view of science, scientific practice, and scientific knowledge. Within the sociology of scientific knowledge, Woolgar (1989b)

11. The interpretive flexibility inherent in the discourse analytic literature has led to a growing interest in questions of "reflexivity," that is, in alternative literary forms that avoid assuming a privileged epistemic position. The clearest exemplar of this concern is Ashmore 1989.

has issued a call for more self-conscious attention to the rhetorical tradition, while Myers (1990) has responded negatively to its more formalist incarnations. Insofar as scientists produce different versions of their social worlds in different social contexts (Mulkay and Gilbert 1982a), scientists' interpretive practices not only produce differing accounts of their interventions into the natural world but also reveal different understandings of what it means to view those interventions as "scientific." This necessarily involves processes of demarcation in which operative communal definitions are negotiated and sustained.

Additionally, it should be noted that the discourse analytic approach frees the analyst (rhetorical or otherwise) from interpretive dependence on the explicitly articulated aims of scientific practitioners. As Mulkay and Gilbert argue, "The goal of the analyst is no longer identical with that of the participant. The task of the analyst is no longer to reconstruct what actually happened . . . but to reflect upon the patterned character of participants' portrayals" (1982b, 313–14).

This is not to suggest that the fact that the discourse was produced by scientists (rather than, say, florists or rhetorical critics) is of no consequence. It is to suggest that the discourse analyst need not be restricted to those conclusions recognized or even acceptable to scientists. This is to say that the discourse analytic tradition is far more "democratic" (some say anarchic) than its predecessors. The domain of appropriate questions to be asked about scienec and scintific practice need no longer be circumscribed by a neo-positivist understanding of science or by any other a priori grant of authority. Indeed, the key to a critical interrogation of science in contemporary culture is precisely the expansion of our methods of interrogation. Recall Fuller's (1989b) admonition that we must stop thinking about science as having some sort of natural integrity which necessarily compels us to interpret it exclusively in its own terms. Altering the terms and methods of our investigations can only provide broader and richer understandings of the epistemic privileges granted "scientific" ways of knowing in modern culture.

As might be expected of a research program which seeks to change the "constitutive questions" of the sociology of scientific knowledge (Mulkay 1981, 172), the discourse analytic program has attracted at least its fair share of criticism. I will bracket the majority of its indictments, however, and will only discuss the issue most directly relevant to a rhetorical study of demarcation. This issue involves the approach's operationalization of the notion of "discourse." Tibbetts and Johnson have argued that an exclusive reliance on discourse as evidence for claims about scientific knowledge is inadequate because of the frequently ambiguous relation between discourse and praxis. They ask,

"Can we restrict ourselves as sociologists of science to discourse if that discourse both defines and is defined by praxis and the ongoing contingencies of the conduct of inquiry?" (1985, 747). The thrust of this criticism appears to be that there is a distinct (presumably nondiscursive) dimension to scientific practice which is systematically ignored in discourse analytic accounts of that practice. In short, discourse analysis is said to be abandoning one overly restrictive view of science (e.g., Mertonian) in favor of a different, but equally restrictive, discursive one.

This objection is simultaneously quite reasonable and quite beside the point. Evaluated on the ability of discourse analysis to meet its own objectives, such a criticism is innocuous. To the extent that discourse is explicitly taken as a sufficient research topic, the fact that it does not account for a variably described "praxis" does not render the program ineffectual. At best, it may call into question the constitutive aims of discourse analysis in the sociology of scientific knowledge, but that is a matter best left for sociologists to debate.

For a rhetorical account of demarcation, this is a criticism of more importance, however. I can easily admit that there is a dimension of scientific practice, for instance, placing *x* wire in the TEA laser, which either is nondiscursive or, at least, is best studied nondiscursively. It is not likely that the state and federal governments routinely approve massive expenditures simply to allow researchers to talk. To concede that scientists do things (other than talk), however, does not deny that a meaningful understanding of that practice remains intrinsically discursive and rhetorical. The standards shaping strategic choices of "nondiscursive" praxis are themselves the products of historically and contextually situated processes of negotiation. As such, our understandings of the technical decisions based upon them rely on their discursive origins. Rather than granting interpretive primacy to the recalcitrances or "resistances" (Pickering and Stephanides 1992), perhaps we might probe the notion of rhetorical recalcitrance. Woolgar seems to have a similar notion in mind when he observes, "The strength of a scientific explanation is no more than its degree of resistance to deconstruction" (1988b, 108).

Second, and more important, one can make more sense of laboratory practices by focusing on those discursive dimensions. Unless we are willing to posit a pristine, noninterpretive realm in which we can unproblematically apprehend said practices and redeem our accounts of them, we cannot sustain the artificial praxis/discourse distinction that Tibbetts and Johnson (1985) trade upon. Indeed, as Mulkay and Gilbert point out, "Observers of laboratory life have emphasized that *behavior* can be observed in the laboratory, but its meaning, its character as social action, is constituted through and inseparable from scientists' accounting practices" (1982b, 313). Insofar as the goal of this story is an account for the demarcation practices of science as social action, a highly (yet not synonymously) discursive understanding of discourse and praxis is demanded.

Discourse analysis assumes a strategic understanding of particular discursive practices. For example, Gilbert and Mulkay performed follow-up interviews with the authors of the formal articles studied, accumulating impressive evidence to suggest that particular adaptations were intended as adaptations to particular contingencies. They noted, for example, that the systematic elimination of aspects of personal involvement was done to "enable the author to avoid responsibility for errors" (Gilbert and Mulkay 1980, 287). More striking, perhaps, is their quotation from one of the authors on his personal reaction to his own purification of experimental practices. The researcher quipped, "It's appalling really, it's taught all the way in school, the notion that you make all these observations in a Darwinian sense. . . . That's just rubbish. . . . God knows, you see everything. And in fact, you see what you want to see for the most part" (287).

While this is compelling evidence of adaptation to a particular social context in specific cases, such adaptation need not be an absolutely conscious and deliberate action. Indeed, if we take seriously the potential of previously negotiated tacit commitments (Polanyi 1958), such rhetorical adaptations might be seen as even more compelling evidence of the practical influence of symbolic demarcation practices. The rhetorical authority of a given vocabulary rises in proportion, it seems, with its "taken-for-grantedness."

What I should make clear here is that there are important linkages to be drawn between particular interests and scientists' discursive practices (Barnes and MacKenzie 1979; Pickering 1989; Yearley 1982). It is around those interests that social (including discursive) practices are organized. Barnes and MacKenzie have argued that when scientific practitioners evaluate "knowledge, the process should be understood to some extent in terms of the goal oriented character of its thought and activity" (1979, 52; see Dolby 1975, 1979; and Frankel 1976). What appears to be lacking is an account of the topoi that are available to scientific rhetors as resources in periods of controversy and of the transformative power of those resources.

While one could not hope to catalog all possible topoi, one such topos is especially compelling. As I noted earlier, "impersonality, detachment, and universality" (Gilbert and Mulkay 1980, 270) are constructed as defining characteristics in a way quite similar to my previous reinterpretation of the Mertonian norms of science. This narrow rhetorical construction of rationality is one in which observation is the product of the "unembroidered evidence of the senses" and extrapolation from observation is guided by clear-cut, unproblematic inferential processes. Accordingly, as Gilbert and Mulkay emphasize, "observational differences and differences of interpretation are seen as being due necessarily to the encroachment of non-scientific or non-technical factors" (1980, 289).

Understood in this way, "errors" are seen as threats to the proper functioning of science. Truth is construed as equivalent to the product of proper scientific procedure. Falsity is construed as equivalent to error, which is a product of some variety of nonscientific practices. Error is not merely something which must be explained; it must be explained away, that is, explained away from the domain of science. While it can be argued that science, as a social practice, is painstaking in its "tracking down" and repudiation of error, when such errors are tracked down, the basis of their repudiation is the characterization of them as arising from practices failing to meet the standards of scientific behavior. Gilbert and Mulkay provide a cogent analysis of such practice and argue that accounts such as "failure to understand, prejudice, insufficient experimental skill, false intuition, personal rivalry, emotional involvement, irrationality and general cussedness" (1982, 176) serve to situate the locus of error outside of science.

At base, this "asymmetrical" process of accounting for "error" is a practical exercise in demarcation, that is, a process of establishing the grounds for scientific inclusion via rhetorical exclusion. Discourse analytic work offers potential insight on demarcation practices such as those emerging from rather obvious controversies, such as the creationist controversy, but also from "intrascientific" disputes such as those surrounding the "discovery" of cold fusion. It is in the formal accounting processes embodied in the research report that the "causes" for experimental error are defined and the boundaries of legitimate scientific practice are demarcated.

While certainly not a dominant strain in the sociology of scientific knowledge, there is work that suggests that the particular boundaries of science are "rhetorical accomplishments" (Woolgar 1988b, 26). While what, precisely, is meant by *rhetorical* in this work is problematic, the story is worth telling in the service of a rhetorical account of demarcation practices. The most extensive development of this general perspective involves scientists' "boundary work," defined as "those acts and processes that create, maintain, and break down boundaries between knowledge units" (Fisher 1990, 98; see Gieryn 1983a, 1983b; Gieryn, Bevins, and Zehr 1985; and Gieryn and Figert 1986, 1990).

Originally articulated by Gieryn (1983a, 1983b) in decidedly Mertonian fashion, the boundary work story has of late flirted with more relativist developments in the sociology of scientific knowledge (Gieryn and Figert 1990) and post-structural theory (Fisher 1990). In either case, Gieryn takes as his guiding problematic the practical necessity and empirical reality of demarcation practices, observing that "even as sociologists and philosophers argue . . . the possibility of demarcating science from non-science, it is . . . routinely accomplished in practical, everyday settings" (1983a, 781). He goes on to highlight the evaluation of National Science Foundation funding applications and the selection of public school textbooks as examples of the practical establishment of relevant demarcation criteria.

This is not to suggest that such cases reveal the *true nature* of science. Indeed, Gieryn is flatly critical of "the apparently futile attempts by scholars to decide what is essential and unique about science" (1983a, 781). His concern is to explicate the processes by which such demarcations are practically accomplished. He recognizes that the considerable epistemic and material advantages available to science and scientists make the grounds for those classifications a topic of considerable social and critical importance. Rejecting Larry Laudan's (1983) gambit, he argues that "so long as scientists refer to pseudo-science and non-science, and they do, of course, the sociologist will face a demarcation problem that is anything but 'pseudo'" (Gieryn 1983b, 60).

On this assumption, Gieryn offers a sociological and historical account of boundary work: "a combination of rhetorical and social organizational devices designed to exclude some people and their knowledge claims from science" (1983b, 60). It is analyzed as a "rhetorical style . . . in which scientists describe science for the public and its political authorities" (1983a, 783).[12] Demarcation practice, understood as boundary work, is a practical activity of scientists which is carried out to distinguish scientists from others who might claim the authority and resources sought (and most often, acquired) by traditional scientific structures. Social-organizational devices, on Gieryn's account, are "institutionalized mechanisms for conferring the status 'scientist' only on those who qualify" (1983b, 68). Rhetorical devices, on the other hand, are understood as the discursive application of criteria by the institutionalized gatekeepers in conferring that status.

This general orientation was used to investigate the boundary work activities of the British "statesman for science," John Tyndall (Gieryn

12. Gieryn's selection of "style" is problematic. As Holmquest (1990) points out, a more appropriate category of such rhetorical practices (at least according to the classical canons of rhetoric) would be "invention," in which an orator discovers, creates, "invents," the materials of discourse.

1983a), and, subsequently, to examine two twentieth-century creationism trials (Gieryn, Bevins, and Zehr 1985). In the former study, Gieryn demonstrated how Tyndall "attributed selected characteristics to science that effectively demarcated it from religion or mechanics, providing a rationale for the superiority of scientists" (1983a, 784). In the study of the professionalization-motivated boundary work in the Scopes trial and *McLean vs. Arkansas,* it was argued that "distinctive images of the relationship between science and religion were used on different historical occasions to advance . . . professional goals" (Gieryn, Bevins, and Zehr 1985, 406).

More recent work has expanded and enriched original discussions of boundary work, including Foucauldian analyses of power (Fisher 1990) and scientific popularization (Gieryn and Figert 1990). Taken together, this literature coheres around one fundamental conclusion, summarized aptly by Gieryn and Figert: "Science has no universal definition, no essential characteristics. . . . The qualities, practices, and accomplishments attributed to science are explicable as rhetorical tools used by those whose practical ends and interests are served by making those attributions" (1990, 90).

Beyond the obvious crucial recognition of the rhetorical dimensions of demarcation (however defined), sociological boundary work offers important conceptual support for more inclusive rhetorics of demarcation. Initially, these stories provide a cogent reconceptualization of the allegedly constitutive normative structure of science as meaningful only in its deployment. Gieryn argues that "the familiar features attributed to science alone, its objective, cumulative, empirical, theoretical, and reliable nature, are . . . boundary setting devices used [by scientists] to demarcate themselves from non-scientists" (1983b, 60). The epistemic base of power for such presumptions is that they are taken as "natural" reflections of science and scientific practice. I would argue that they are variably deployed in order to fulfill particular rhetorical ends, that is, to serve particular goals in particular rhetorical situations.

A second important implication follows from Gieryn's conceptualization of boundary work as intrinsically contextual. In other words, subtly different demarcation criteria are advanced in order to sustain boundaries between different social and epistemic competitors. He argues that the potential inconsistency of various proposed demarcations ceases to be an analytic problem when one recognizes that such demarcations might well have been intended to exclude different sets of competing social practices. For example, his study of Tyndall's nineteenth-century boundary work reveals that "the empirical and useful fact was the keystone of science as *not religion* but the abstract and pure theory as the cornerstone of science as *not technology*" (1983a, 792). There is no principled reason to expect that the demarcation discourses of scientists reveal any particular consistency. Indeed, there is considerable reason to suspect that the particular configuration of interests which constrains those discourses would alter the constitutive nature of any symbolic construction of science.

A final implication of Gieryn's formulation warranting mention here is that it is cast at a level of generality that allows it to be applied in cases other than those in which the constructed demarcation is between science and pseudoscience. Given that the guiding imperative underlying boundary work is the protection of particular resources, it is reasonable to expect that such strategies could be found in cases of intrascientific controversy as well. Gieryn speculates that "the same rhetorical style is no doubt useful for ideological demarcations of disciplines, specialties, or theoretical orientations within science" (1983a, 792). This seems to me an especially useful and revealing insight. While demarcation practices might be seen as more relevant to the question of science versus nonscience, the story need not end there; demarcation is practically managed even in cases of controversy within what is generally agreed to be science. The cold fusion case study will show that the boundary work stemming from resource allocation struggles which characterize traditional disputes between, for example, the sciences and the humanities, are not restricted to such disputes. It will reveal that intrascientific controversy is also an exercise in interest-based communal definition and rhetorical struggles over the relative legitimation of those definitions.

Sociological conceptions of boundary work are not without their limitations, however. While similar in theoretical interest, our formulations differ in fundamental ways, speaking to different audiences and from quite different disciplinary goals. The point of this discussion is simply to show the potentially fruitful sociological grounding of any critical rhetorical approach to scientific texts.

A particularly significant limitation of sociological accounts is that they do not or cannot provide any meaningful reflection of the texture of meanings in any given text. There is an implicit assumption that given "objectives" map unproblematically onto the discourses of boundary work. Consider, for example, Gieryn and Figert's discussion of mediated portrayals of Richard Feynman's "o-ring" experiment before the Rogers Commission hearings on the causes of the Challenger disaster. Reflecting on journalists' constructions of the event as pivotal, they observe that "reporters . . . construct science and mercilessly exploit scientists [like Feynman] to give their stories credibility" (1990, 90–91). While there is some discussion of journalistic routines as inventional constraints on this so-called exploitation, the discourses of journalists appear to be oddly detached from broader social constituencies.[13]

Ironically, this detachment is also evinced in the discussion of "rhetorical" and "social-institutional" devices in boundary work. Most notably, the distinction succumbs to the temptation to portray rhetoric as simply window dressing, while institutional factors fundamentally control which windows get dressed. A rhetorical account of demarcation suggests that the very status of institutional gatekeeper is itself a rhetorical designation, hammered out in the crucible of the ongoing rhetorical practices of science.

In addition to failing to demonstrate rhetorical sensitivity to discourse, such an assumption could lead to an unjustifiably idiosyncratic understanding of demarcation practices. What seems to be missing from this account is the constraint reciprocally exerted by broader hierarchical structures on the production and legitimation of particular boundary discourses. As it is, boundary work's implicit view of lone scientific advocates advancing their own demarcation criteria apart from those structures is a curiously asocial sociological perspective.

Finally, Gieryn's account is apparently unable to talk, in any meaningful way, about the dialectical nature of demarcation. He does indicate that particular strategies of boundary work are adapted to different epistemic competitors. This analysis, however, has not included an account of the adaptive influence of those competing discourses. To the extent that rhetoric is controversial discourse, I think much remains unsaid about the intersubjective negotiation of demarcation standards between and among interest-driven social actors (scientists, legislators, journalists, etc.). While I risk judging too harshly, it could be argued that Gieryn implies that there exists some static corpus of boundary work "stories" which are deployed without regard for the particularities of the discourses to which they ostensibly adapt.

I think the demarcation story is best told without such encumbrances; rhetorical demarcation practices are both rhetorically and historically adaptive. The creationism case study will demonstrate that the traditional scientific community's response was constrained to adapt to the changing nature of the creationist alternative from the Scopes era to the era of the Institute for Creation Research. As Cooter has argued, "The study of pseudo-science has value . . . for the sake of revealing more about the science that negotiationally defines pseudo-science through its interactions with it" (1980, 260).

13. Fisher's (1990) discussion of boundary work as an instance of Foucault's power/knowledge in action avoids this unnecessary "individualization."

Historicizing Science and Social Studies of Science

The foregoing analysis has indicated that the "boundaries" of science, far from being the timeless and transcendent structures envisioned by traditional philosophies and sociologies of science, are best understood as symbolic constructions that are advanced and sustained in particular contexts by particular sets of social actors. By implication, there is a strong historical imperative underlying the current understanding of demarcation practices, entailing some degree of historical variation. Of course, the vast body of literature in the history of science provides precisely this sort of evidence (Giere 1977; Freudenthal 1988; McAllister 1986). Anything resembling a complete review of this literature would extend far beyond the concerns of this study. Accordingly, my strategy here is to appropriate a limited number of studies in order to illustrate dominant themes. Subsequently, I will take up those historical studies concerned most straightforwardly with demarcation.

Philosophical Questions and Historical Answers

In her study of canonical histories of science from the seventeenth through the early twentieth centuries, Rachel Laudan (1993) suggests that they serve as apologias for science, seeking to deflect public critique or, at least, linking the conduct of historical research to the services of science. In some ways, I think, this connectedness is echoed in recent historical research, although now the historical work seeks not necessarily to celebrate science (though it can have that effect), but to provide evidence that either supports or challenges philosophical claims.

As the locus of those claims, then, philosophy appears to function as the intellectual catalyst. Consider, for example, Chen and Barker's 1992 study of David Brewster, Henry Brougham, and the emission-undulatory dispute of the 1850s. Challenging dominant themes in the sociology of scientific knowledge that, as they see it, marginalize individual agency and its corresponding focus on cognitive explanations of scientific practice, Chen and Barker conclude that "the behaviors of these principals do not mesh with the Lakatosian view of scientific closure, as enacted according to verification of excess empirical content" (1992, 76). Similarly, Corry's 1993 study of Kuhnian revolutions in the history of mathematics is organized around similarly philosophical heuristics, though with an eye toward contributing to more nuanced historical understandings. He describes his goal: "Rather than discussing whether a definition of scientific revolutions can be advanced such that revolutions are found to have occurred in mathematics [a trajectory that would give historical motives primacy], it is asked whether by analyzing the history of mathematics in Kuhnian terms, new insights are attained which would otherwise have been overlooked by historians and philosophers of science" (1993, 343).

My point here is not that historians or anyone else should cling tenaciously to the patently artificial boundaries we construct to protect our institutional turf within academe. Rather, it is to suggest that utilizing close historical studies to provide richly textured evidence for philosophical accounts of science positions those accounts as fundamental. Such positioning, I think, detracts from the effort to come to grips with the heterogenous symbolicity of scientific practice, an effort that challenges the necessity, even utility, of canonical philosophical accounts of what science is or ought to be.

A second theme I want to highlight here involves the important attempt to reveal how the tension between individual and institutional agency is worked out in particular historical episodes, a topic on which Chen and Barker touched. At this level, historical studies of science dovetail nicely with rhetorical accounts, insofar as the latter proceed generally from the assumption that rhetoric requires rhetors, whether at the individual or the institutional level. In characterizing the outcome (or, at least, the goal) of his studies of Newton's experiments with light during the 1660s and Gray's electricity research in the early eighteenth century, Ben-Chaim insisted that "this study re-introduces individual agency to the study of science" (1992, 535). Resisting the sort of epideictic language described earlier, Ben-Chaim argues that "to construe their achievements as contributions to a collective endeavor termed 'science' is to legitimize science rather than to critically understand their conduct" (1992, 555).

Similarly, De La Bruheze's study of the development of radiological weaponry during World War II portrays the canonical internal versus external dichotomy as enacted as a consequence of individual actors' judgments. He maintains that "whatever the actor's strategies, separate internal and external trajectories developed, based upon the different positions, commitments, interests and objectives of the actors" (1992, 226). Here, I think, one gets a clear sense that the choices made, actions taken, and implications of local actors are more than more or less appropriate reflections of real scientific behaviors. They are, it seems to me, better understood as the historically situated practices of actors in response to the situational exigencies impinging upon them, both as individuals and as representatives of institutional structures. Such "rhetors" (when engaging in symbolic practices) need not be seen as opposed to institutional structures. Indeed, as Althusser has made clear, we are "always already" interpellated into those structures and draw, to a large extent, our identities from those patterns of interpellation.

This tension between levels of analysis is sustained in much of the historical work concerned most directly with acts of demarcation, that is, with those institutionally embedded individual practices that function to mark off the domain of science from other sets of practices. In his impressive survey of the history of science and its implications for recent trends in the sociology of scientific knowledge, Shapin argued that the history of science has rather directly dealt with demarcation issues. The historical approach is characterized by a commitment to "trying to ascertain how historical actors *themselves* defined what belonged to science [or natural philosophy, or whatever term and cultural domain was indicated] and what did not" (1982, 177; see Lepenies 1977). A very rich body of literature cogently demonstrates that the nature and cultural position of "science" in the past bore little resemblance to its current intellectual progeny (Bayertz 1985; Brush 1983; Cale 1972; Cooter 1976; Crombie 1980; Daston 1978; Eckert 1987; Kirsch 1980; Moss 1989; Olson 1982; Oreskes 1988; Palfreman 1979; Roszak 1976; Shapin 1979, 1980, 1984; Shweber 1986; Van Den Daele 1977; Wallace 1989; Yeo 1986; R. Young 1973; Zappen 1986).

Perhaps the dominant historical demarcation which has been drawn is that which purports to distinguish science from religion (Dupree 1986; Kirsch 1980; Numbers 1985; Westfall 1986b; Westman 1986). As particular ecclesiastical hierarchies lost, gained, and struggled for moral and cultural power, particular understandings of the natural world and the appropriate means for its investigation and control were variably established. While I will elaborate no further here, it will become clear that such demarcation struggles were, and remain, at base struggles for the primacy of spiritual over traditionally secular ways of knowing.

A second important line of demarcation with particular relevance to this study concerns the progressive professionalization of science and scientific practice (Abir-Am 1987; Allison 1979; Basalla 1976; Bunders and Whitley 1985; Cambrosio and Keating 1983; Franck 1979; Gieryn, Bevins, and Zehr 1985; Kohler 1982; Kuklick 1980; Lankford 1981; O'Connor and Meadows 1976; Pinch and Collins 1979; F. Turner 1978; Westrum 1977, 1978; Weyant 1980; Whitley 1977). This literature does not suggest that there was a self-conscious struggle either to include or to exclude from science those who had little or no professional training. The primary concern was with demarcating the legitimate spheres of expertise among and between particular specializations. As Shapin noted, "By the late nineteenth and early twentieth century, the boundaries between the professional scientific community and mere amateurs had been fairly well defined in most areas of science. However, there were specialties in which the professional-amateur demarcation was not as rigid as it had become elsewhere" (1982, 173). Lankford's (1981) discussion of the authority struggles in astronomy is an exemplary illustration of this sort of demarcation practice.

The implications of these historical narratives for our attempt to understand demarcation rhetorically are clear. Taken as a whole, they suggest that questions regarding the constitutive nature of scientific practice are implicated in interdisciplinary controversy. In cases where issues of the most appropriate, hence most scientific, set of experimental practices are negotiated, a rhetorical account of demarcation will illuminate important processes of communal (re)identification. The cold fusion case study in chapter 6 illustrates the presence and implications of such processes.

As I have noted, individual historical studies are not as important for my present concerns as are their general implications. Most fundamentally, they strongly reject any ahistorical, decontextual understanding of scientific practice. While certainly not an idiosyncratic process, the historical evolution of science is not delimited by timeless boundaries. As Shapin observed, "In the making of scientific knowledge *any* perceived pattern or organized system in nature, in culture, or society may be pressed into service. These patterns are resources for understanding the natural phenomena in question" (1982, 178). Accordingly, the rather undeniable success (and utility) of "science" is due not to a set of timeless characteristics, but rather to the flexibility of its constitutive aims and practices.

This contextual understanding of the variability of demarcation practices leaves unanswered, however, the question of the influences on any particular act of demarcation. It is one thing to say that science has been (and will be) demarcated in various ways at various times. It is quite another to demonstrate the contextual constraints which influence any particular act of demarcation and the locally embedded discursive dynamics of those acts.

Such a demonstration appears to be the direction in which Cunningham's historical study of demarcation appears headed, though it falls prey ultimately to the individual versus institutional dichotomy noted earlier. Spurred by the historian's topical quandary "Are we studying the right subject?" he concludes that "we obviously need to be in possession of criteria which enable us to identify science in the past correctly every time" (1988, 365). The root of the difficulty, as Cunningham sees it, is the historian's enactment of two assumptions. The first takes for granted the "specialness" of science, while the second turns on the historian's inevitable "present-centeredness," the tendency to interpret the past through contemporary lenses. In general, his remedy is to insist that "science is a human activity, wholly a human activity, and nothing but a human activity" (370).

While seemingly unexceptionable on its face, the implications of Cunningham's formulation bear close scrutiny. First, such a position excludes (at least from historical study) the material objects with which human actors actually interact in the doing of science. I do not mean to suggest here that bubble chambers are quintessential humans, less yet rhetors. Nor do I feel any compelling urge to abandon the human/nonhuman distinction altogether in favor of Latourian "actants." Nonetheless, such a rigid bifurcation belies the variability with which historical configurations can draw upon myriad constituents in their moments of stabilization, moments that enact science.

A second limitation proceeds from Cunningham's insistence that not only must science be seen as solely a human practice, but an intentional human practice as well. Building on a quasi-Wittgensteinian language game, he insists that "it is . . . clear that at a given moment a person is either doing science, or he is not; it is something which, if he is doing it, he intended to do" (1988, 377). Beyond committing what Fuller (1988) labels the "intentional fallacy," this view gives rise to troubling methodological implications, most notably an exclusive reliance on "insider" discourse in attempts to reconstruct the historical configurations that the historian characterizes as science. As Cunningham puts it, "This means that to describe them as having been engaged in science can only be a true description of their activity if it could have been *their own* description of their activity" (1988, 378).

A rhetorical view of science, especially one concerned with the permeability of boundaries, seems ill served by such a dictum. In the first instance, it takes as given the boundaries of science, even if "given" discursively, a sort of rhetorico-scientific fundamentalism. Of course, whose discourse we take as reflecting most accurately her or his perceptions of doing science remains utterly opaque, and our own interpretive practices are said to be incapable of offering any insight on the matter. This, I'm afraid, reduces such studies (whether historical, sociological, or rhetorical) to banal exercises in stenography.

This view, then, carries the related assumption that science, whatever it is, can be described, indeed practiced, only by those voices that we judge scientific. Such a methodological assumption, of course, carries with it the equally troubling implication that marginalized voices, those that articulate nontraditional conceptions of science, or simply those who offer new ways of talking about old things, will remain marginalized. This seems highly unfortunate in that it denies the promising utility of an alienated reading strategy. Such a strategy might allow us to interrogate, rather than to simply describe as natural, precisely those cultural configurations around which centuries of border clashes between "science" and "nonscience" have been waged. By including previously excluded "voices," we can begin to move in the direction of more constructive analyses of science and its constituent, and constitutive, rhetorics.

While demarcation is hardly its primary focus, Shapin and Schaffer's monumental *Leviathan and the Air-pump* (1985) might be read as the most insightful endorsement of the constructedness of science and its boundaries in the historical literature, and so read without the methodological dead ends attendant to Cunningham's formulation. As they put it, "What we cannot do if we want to be serious about the historical nature of our inquiry is to use such actors' speech unthinkingly as an explanatory resource. The language that transports politics outside of science is precisely what we need to understand" (341–42).

Their study of Hobbes and Boyle illuminated what Latour described as the "co-production of science and its social context" (1990, 147). They maintain that the key motivation for historical studies (one rhetorical critics would be wise to share) should be to understand how the local actors arrayed multiple constituents "with respect to *their* boundaries . . . and how, as a matter of record, they behaved with respect to the items thus allocated. Nor should we take any one system of boundaries as belonging self-evidently to the thing that is called 'science'" (Shapin and Schaffer 1985, 342).

Calling into question the very analytic categories upon which almost all extant historical and sociological accounts of science trade, however unreflectively, ("scientific" and "social"), Shapin and Schaffer's analysis moves forcefully in the direction of the permeability that is presumed in a rhetorical account of demarcation. Latour's praise is effusive on this general issue, though cast in quite different terms. He concludes that "others have studied the practice of science; others have studied the . . . context of science; but none so far have been able to do the two at once" (1990, 151).

Shapin and Schaffer's provocative contention that the range of items "allocated" by particular historical actors in their constructions of science ought not be limited by our invoking a priori standards of "scientificness" has important implications for the ways in which we can understand and enact the "scientific ecosystem" metaphor I developed in chapter 1. In my survey of the sociology of scientific knowledge, I described the various sorts of laboratory activities (heretofore considered "internal") that have been illuminated by the turn to "practice." Here, I want to speculate on the sorts of activities that lie beyond the now shopworn boundary between internal and external constituents.

Perhaps the most central "items" allocated to the practice of science in the contemporary age are the material resources without which "laboratory life" would cease, as Latour has noted. Webster argues that "science and technology are firmly located within the political arena because of their central concern to the state: all governments today are involved in the large scale funding, management, and regulation of science and technology" (1991, 5).

Indeed, Pinch (1990) has argued that the interconnection between funding sources (governmental, academic, or industrial) and practicing scientists is so complete that, in normal activities, the relationship is effaced. It is at those times when normalcy is disrupted that the relationships are brought to light. In his terms, "That science is dependent upon the goodwill and resources given to it by the wider society is . . . made most clear when the representatives of that wider society threaten to turn off the tap" (1990, 220). While I eschew the easy reliance on traditional categories (science vs. wider society), I think this observation warrants consideration. At the very least, it would seem to entail that any meaningful construction of science as practice must incorporate "items" that for too long we have considered as "outside" the domain of real science. This is no mere terminological reorientation. These interconnections are among the practices of science.

This connection raises provocative possibilities, along with its share of conceptual difficulties. Most notable among these is to find new ways of characterizing the configurations of science that once relied on oppositions, but now gain identity through the recognition of connectedness. Cozzens, in her study of the relationship of funding and knowledge growth, highlighted the difficulty of mapping that relationship. She queried, "Is the language of input-output analysis appropriate; and if not, what other language can we develop . . . ? How can we characterize the effects of decisions about funding programmes as they reverberate into the various levels of the scientific community?" (1986, 15).

The case studies of creationism and cold fusion suggest to me that the ultimate indeterminacy of particular configurations renders the quite determinate language of inputs and outputs suspect, at best. The language of the ecosystem, on the other hand, highlights connectedness yet does not presume in advance the particular contours of connectedness, or the particular constituents whose relationships will reverberate most profoundly.

In any case, the key conceptual shift here must be away from viewing items such as funding programs as likely to pollute *real* science, and toward embracing a more heterogeneous and inclusive vocabulary. Of course, such a strategy is politically loaded. As Jasanoff noted, "Scientists have been quite successful in protecting this claim of exclusivity, jealously guarding their power to define the public image of science, and warding off competing claims by rival disciplines" (1987, 196).

Nonetheless, this conceptual and rhetorical reorientation in the ways we understand science and demarcation need not be construed as emerging irrationally in unprincipled fashion from scientific or nonscientific prejudice. While they may do so, demarcation practices are more meaningfully understood as the symbolic construction of a domain in which privileged social, technical, cognitive, and/or professional interests are preserved through the legitimation of a particular authorizing community. As Nowotny maintains, "To question the motivations of scientists is not to belittle their genuine contributions, but simply to show that they too have interests that may be at stake" (1979, 7). Similarly, Collins and Yearley maintain that "making science a continuous part with the rest of our culture should make us less intimidated and more ready to appreciate its beauty and accomplishments. It should make us more ready to use it for what it is, to value its insights and wisdom within rather than without the political and cultural process" (1992, 309). Hence, a rhetorical account of demarcation functions simultaneously to demystify and challenge the unreflective epistemic privilege accorded science and more clearly to elucidate the standards by which everyday scientific practice is evaluated. It is that account that is developed and enacted in the next three chapters.

4 DEMARCATING SCIENCE RHETORICALLY

These should be heady days for those scholars interested, as a matter of disciplinary affiliation, in mapping the discursive dynamics of scientific practice. As such a scholar, I suppose I should have known better. I had just begin to allow myself to hope (if not believe) that the various rhetorical turns reconfiguring intellectual geometry in recent decades would move rhetorical studies, if not to the center, at least a bit further from its accustomed position at the institutional margins. But no. Our own well-honed hand-wringing ushered in an essential, very valuable, but still damnably deflating reflexive moment. Dilip Gaonkar, for example, has argued that "in its current form, rhetoric, as a language of criticism, is so thin . . . that it commands little sustained attention" (1993, 263). Similarly, Alan Gross (speaking, I presume, as something of a hostile witness on this account) left open the apocalyptic possibility that "as a discipline, rhetoric of science is intellectually superfluous" (1994, 6).

By any standard, I think that rhetorical critics have contributed productively to the interdisciplinary discussion of the rhetorics of science and inquiry. The essays in two edited collections, for example, provide ample evidence of this contribution: *The Rhetoric of the Human Sciences* (1987), edited by Nelson, Megill, and McCloskey; and *Rhetoric in the Human Sciences* (1989), edited by Herbert Simons. Whatever the genesis of this disciplinary psychosis, we must now move ahead and confront the necessity to map promising research trajectories, without assuming a priori that they and we will fail.

As with most meaningful attempts to come to grips with important ideas, the rhetoric of science literature is rather divided against itself. On the one hand, we celebrate as our critical practices exact the "rhetorician's revenge," exorcising a bit too loudly the demons of the Vienna Circle by standing with Gross to assert that "science is rhetorical, without remainder" (1990a, 213). On the other, we simultaneously retreat, conceding with considerably less gusto that we are apt to be dismissed by those scholars who, as Gross himself puts it, do the hard work of "mastering real science" (1994, 5). In other words, we are reminded to be careful of what we wish for—because we might just get it.

I think this ambivalence, however, stems less from globalizing rhetoric to the point of meaninglessness (Gaonkar 1993) than from a far less reflexively monitored impulse—globalizing science. Even as rhetorical critics trumpet the fall of the multiple "isms" suggested to constitute *the* scientific method, our reading strategies consistently avoid anything resembling oppositional, or even alternative, characterizations, even to the point at which we premise our critiques on formulations of "real science." A more productive orientation, I think, would engage the implications of our work for what canonical histories portray as our raison d'être: the polis. I short, what I mean to suggest is that one potentially valuable contribution of a rhetorical account of science generally, and of demarcation in particular, is the crafting of more meaningful conceptions of *phronesis* in the postmodern age. Of course, that move presumes a much broader conception of the polis than traditionally has held sway, requiring alternate understandings of the shifting natures of public discourse (McGee 1990) and the critical practices impinging on them (McKerrow 1989). A view of science as ecosystem has the potential, I believe, not for re-creating chimerical unity among the postmodern discursive debris, but for enabling fruitful patterns for understanding and adapting to it in the service of community. Demarcating science rhetorically might help prefigure more democratic discourses of and about science in our age.

The "Character" of the Scientific Community

Rhetoric's traditional identification as public, communal discourse suggests that a rhetorical perspective on demarcation must begin with some conception of science as a communal enterprise. This is hardly a novel insight (e.g., Hagstrom 1965). In large part, though, what we make of this conception is the critical (and, too often, underacknowledged) contribution that rhetorical studies makes to the swirl of stories of and about science.

As Aristotle (1991), as well as Perelman and Olbrechts-Tyteca (1969), have demonstrated voluminously, a fundamental characteristic of rhetoric is that it is addressed discourse. That is, it is concerned with the persuasion (intentional or otherwise) of an audience. Using this intuitive commitment as starting point, Overington (1977) recasts our understanding of science from a discrete, asocial enterprise to one in which communal standards of appropriateness and reasonableness constrain the production and legitimation of knowledge claims. Assuming "communitarian practice as the context for the production of scientific knowledge," he argues that "science as a knowledge producing activity may be treated rhetorically *because* the construction of scientific knowledge involves argumentation before an *audience*" (Overington 1977, 143-44). He gives a decidedly rhetorical twist to Polanyi (1958) and Ziman (1968), arguing that "from a rhetorical point of view . . . one could understand the process of constructing scientific knowledge as a way of speaking about specific experiences before a limited and specially trained audience that is authorized to establish that discourse as knowledge" (Overington 1977, 143–44).

Several implications of this perspective merit attention here. First, Overington does not exclude the domain of material practice from his conception of science, acknowledging that the function of the scientific community qua audience is to pass judgment on a particular petitioner's account of his or her experimental or observational practices. This aspect of Overington's analysis would seem to exempt it from the occasionally scathing critiques directed at certain rhetorical perspectives which seem to imply that there is no meaningful empirical dimension of science. Of course, this does not imply that a given scientist's apprehension of her or his experiences is not mediated by processes of interpretation (Delia and Grossberg 1977; Hanson 1958).

Overington also insists on the unique nature of the scientific audience. Insofar as this audience is conceptualized as "limited and specially trained," Overington does not sanction a view of science as simply another form of life. He argues that "to speak as a scientist is a privilege, not a right. Many people use scientific terms, but they are not speaking scientifically; to do that requires that one be recognized by other scientists as one of themselves" (1977, 155). The standards upon which that recognition is predicated form the substance of demarcation practices and the focus of this study.

In general, Overington offers an insightful and convincing analysis of the audience-oriented dimension of scientific practice. There are certain problematic assumptions, however, which would be clarified by a rhetorical investigation of demarcation practices. Overington's analysis does not explicitly take up the issue of demarcation. That, of course, is not itself an indictment, but this oversight does lead to conceptual confusion. To suggest that science is a community of scientists which holds authority to commission new scientists is, in many ways, little more than a tautology. Overington's analysis of audience as authority might easily (and of theoretical necessity) be extended beyond his current concern with the authorization of knowledge claims advanced by those who meet the definition of *scientist* to a deeper consideration of the rhetorical processes in which those very definitions are authorized. In effect, the nature of science is taken for granted; to borrow Latour's (1987) term, it is "black boxed." A rhetorical study of demarcation promises to pry open that conceptual black box.

A similar limitation of Overington's analysis involves the lack of clarity regarding the standards which are used to test competing knowledge claims. While limitations of space might partially explain this omission, it is important to note that such standards are typically taken to be themselves unproblematic and not suitable for rhetorical analysis. It is only the *application* of these standards in particular contexts that has been the object of attention. Rhetorical scholars should guard against an unreflective reification of the standards themselves. In Overington's discussion of the "long-term" authorization of knowledge claims (1977, 156–57), there seems to be an implicit assumption that the standards for evaluation remain static amid the passage of time. Such need not be the case, however, (e.g., Kuhn 1962; Pearce and Rantala 1984; Toulmin 1972). Dramatic scientific innovations, such as chaos theory, might well implicate a radically novel set of appropriate standards (Keith and Zagacki 1992). Insofar as Overington's standards are taken to be intrinsically necessary for scientific authorization, a more focused investigation of those constitutive standards would be tantamount to an exercise in demarcation. It seems plausible that scientists not only argue *from* these standards; they also (at least implicitly) argue *about* them in particular contexts. In doing so, they implicitly construct the nature of science and scientific knowledge. As Bazerman puts it, a community "constitutes itself in developing its modes of regular discourse" (1987, 129).

To say simply that science is best understood as a communal audience that stands in judgment of its constituent members, however, is to reduce a complex social phenomenon to a rhetorical matinee idol, superficially attractive perhaps, but woefully lacking in development. Among the issues most lacking is the question of the particular contexts in which the boundaries of this "audience" are drawn. As a consequence, I want now to take up the rhetoric of science literature regarding the contextual natures of science. Most generally, context is operationalized in one of two ways. First, studies articulating an internalist conception of context exploit a view of scientific practice as fundamentally a matter of argumentation conducted *within* the context established *by* a prevailing authoritative community. Externalist conceptions of context, on the other hand, serve to probe the relative influence of the larger sociocultural milieu on the nature and functions of science.

In telling this part of the story, I consciously exploit a patently artificial narrative convention by demarcating the internal and external conceptions of "context" operating in the literature. The irony is wholly intentional. Indeed, I will ultimately try to hoist that very convention on its own petard in calling for a rhetorically constituted series of boundaries between these most fundamental categories.

Perhaps the most extensive rhetorical treatment of science as an internally contextualized discourse community is Prelli's *A Rhetoric of Science: Inventing Scientific Discourse* (1989). This study borrows heavily from classical inventional theories of Aristotle and Cicero (by way of Richard McKeon and others) to construct a view of scientific discourse that is "technically sound as a theory of rhetoric and is representative of scientists' practices when they engage in discursive activity" (Prelli 1989b, 6). Prelli's intention is to reveal the heretofore masked rhetorical dimensions of "doing science" (8).

In one sense, Prelli's enterprise is descriptive; he describes "constructs and categories that derive from actual discursive practice in science . . . and [provides] formulations that explain both how discourse is made and how it is judged *as science*" (1989b, 8). The inventional process in science is accomplished by arraying the classical stasis categories (conjectural, definitional, qualitative, and translative) with the superior stases of science (evidential, interpretive, evaluative, and methodological). The multiple interfaces between those points of controversy yield three broad classes of topoi (problem-solution, evaluative, and exemplary) that can be used to attack or defend the reasonability of scientific claims. Thus, Prelli argues that science is rhetorical most fundamentally because it "involves selective use of symbols to induce cooperative acts and attitudes regarding a mediating symbolic orientation" (87).

The utility of this methodology is then demonstrated in case studies of several scientific controversies, such as memory transfer, solar neutrinos, and ape language acquisition. Taken together, they situate the character of science in the acts of scientists marshaling rhetorical resources to enhance their own credibility and that of their knowledge claims in order to gain assent from the authorizing community of scientific practitioners.

In another sense, however, this view of "setting" is normative in character. Maneuvering between the extremes of long-discredited positivisms and more recent anarchisms, Prelli maintains that science, above all, is *reasonable* rhetoric and that (real) scientists are reasonable rhetors. Relative judgments of reasonability are said to emerge from the scientific community, which warrants the credibility of scientists and the knowledge claims they advance in relation to particular rhetorical "ends." As Prelli describes it, "To be judged reasonable and persuasive in any specific situation, scientific discourse must be perceived as identifying, modifying, or solving problems that bear on a specific community's maintenance and expansion of their comprehension of the natural order" (1989b, 122).

To a degree, I share Prelli's epistemological caution; after all, I concede willingly that the computer screen before me is working, suggesting that someone had it pretty much right in its conceptualization and design. This notion of reasonability, however, seems to replace Mertonian functionalism with its rhetorical counterpart. Construing the impact of internal contextualization in this way seems to entail a noncritical illumination of how well science does what it does, a sort of rhetorical apologia for science. Now, one does not want to deny the apparently undeniable—that science has done its work quite successfully. Limiting the domain of the reasonable to the reduction of ambiguity and increased comprehension of natural order, however, minimizes the constructedness of all three: ambiguity, comprehension, and natural order.

Science, as a rhetorical enterprise, certainly features persuasive appeals regarding natural order being addressed to internal authorizing audiences. Nonetheless, to suggest that only those sorts of appeals merit the status of reasonability portrays science as an asocial practice, remarkably unconstrained by influences that we generally take (often unconsciously) to be nonscientific. Prelli himself makes reference to "the special culture of scientific communities" (1989b, 236) and to the "logic specially valued by scientists" (144), implying a too-simple demarcation between the scientific and un-or nonscientific cultures, not to mention the shifting symbolic grounds for those demarcations. Similarly, consider his discussion of rhetorical "purposing" (esp. chaps. 3, 7, and 10). He suggests that scientific rhetors consciously adopt particular rhetorical strategies as a means to reach the narrowly defined end of scientific reasonability. This focus on conscious strategy implies that scientists' inventional choices are constrained only by their relative potential to redeem knowledge claims.[1] Those choices are subsequently portrayed as rather pristinely unaffected by the more mundane considerations that affect the inventional choices of nonscientific rhetors.

At the very least, this underestimates the ideological functions of science in contemporary culture. It also raises the question of whether or not such a rhetoric of science could possibly achieve the goal of being

1. Beyond constructing an artifice of insularity, this focus on intention also commits the "intentional fallacy," implying that scientists' intentions map isomorphically onto their discourse.

"representative of scientists' practices" (Prelli 1989b, 6). By this, I want to suggest that such internalist accounts of scientific invention ignore the powerful constraints exerted by broader systems and structures of influence on the conduct of modern techno-science (Aronowitz 1988a; Dickson 1988). Put simply, I think such internal contextualizations cast their definitional nets too narrowly. In an attempt to contextualize scientific rhetoric, they seem to distance it from the broader cultures in which it is embedded and from which it draws its material sustenance (Myers 1990; Pickering 1992).

The assumption of a fairly stable, albeit rhetorically flexible, stock of reasonable topoi which stems from this internalist account is also problematic. Most notably, it shortchanges the presence of considerable controversy within science, even as to the very nature of the scientific enterprise (e.g., Bohme 1979; Feyerabend 1979) or to the nature of the natural order that it is supposed to understand (H. Collins 1985). While I do not want to argue that practicing physicists, biologists, and so on spend a great deal of time debating the philosophical parameters of science, I will suggest that there is sufficient disagreement to render Prelli's contention suspect. Indeed, the creationist case study that Prelli advances appears to disconfirm his own position. As I will argue shortly, that controversy features, in some sense, two groups with fundamentally different conceptions of science, quite apart from the more limited (but more politically volatile) question of religion in schools (e.g., Marsden 1980; Taylor and Condit 1988). At base, then, internalist accounts of science exaggerate the homogeneity of science in order to provide a common ground of shared values, which serve as the enthymematic premises for argumentation in scientific controversy. While such commonality might exist, it does so not on the basis of its intrinsic nature, but as a reflection of its rhetorical force.

It is perhaps unfair to criticize internalist work such as Prelli's because it does not tell the story I want to tell in quite the way I want it told. That is not my intent, however. I discuss *A Rhetoric of Science* at such length because I think it demonstrates, by its virtues, the lacunae that I fear await internalist conceptions of the scientific community.[2]

Just as actors' movements are artificially constrained by too-small performance spaces, so a rhetorical account of demarcation is hindered

2. It would be foolish, however, to suggest that some exigencies are not more strictly internalist than others. I take exception to the general characterization of the scientific community as insulated from the vagaries of the social, political, economic, and so on. Keith and Zagacki (1992), for example, aptly describe the rhetorical strategies incumbent upon defenders of "revolutions" in science (e.g., chaos theory) when they are arguing before their peers. Similarly, Lyne and Howe (1986, 1990) have documented the rhetorical trans-

by too small a potential range of discursive contexts. While I do not challenge the conception of science as a community, I echo Latour (1987) in suggesting that the boundaries of that community are not defined as laboratories and journal offices. To broaden the setting and, hence, the story we are able to tell, we need to read science more vigorously into the social, which, in turn, allows us to reread the social into the scientific to the point that the two analytic categories are revealed as precisely that, *our* (not nature's) analytic categories.[3]

We can, however, craft alternate understandings in such a way that the operative contexts in which "science" is embedded potentially include broad sociocultural formations. To extend the ecosystem metaphor here in what I hope are heuristic and productive directions, we can consider the myriad cultural settings of science. Not surprisingly, given the broad cultural strokes with which context is portrayed, the best of this literature tends to be concerned with exemplary cases from the history of science.

John Campbell's important series of essays on the rhetoric of Darwin and the conception and reception of Darwinian evolution (1974, 1975, 1986, 1987, 1989) is perhaps the standard of this sort of historicocritical research. While Campbell's work spans a number of years and substantive topics, it coalesces around an implicit theme of science as reciprocally related to, yet operationally distinct from, the larger social milieu in which it is embedded.

In his account of Darwin as scientific polemicist, Campbell rejects the mythic image of Darwin as the dispassionate and detached investigator and the notion that the "heat of the controversy [w]as the unfortunate by-product of his popularization at the hands of others" (1975, 376). Central to his characterization of Darwin as an effective rhetorical advocate is Campbell's insistence that "against the vastness of human ignorance, Darwin makes an eloquent and moving plea for the just claims of limited knowledge" (390). Campbell subsequently argues that this conception of "limited knowledge" was the moral imperative constraining Darwin's selection of rhetorical strategies.

This insight is relevant to the current project insofar as it is one of the earliest references (in the rhetoric of science literature) to a particular

formations of "punctuated equilibria" and "sociobiology," as those discourses have moved among and between alternative audiences, or "discourse frames" (1990, 136–47). Ceccarelli (1994) also provides compelling evidence of the influence of alternative audiences and goals on and in intradisciplinary scientific texts.

3. This is not to diminish the value, indeed necessity, of those (or any other) analytic categories to interpretive research. It is simply to underscore the need for reflexivity and for recognizing the ultimate indeterminacy of all such categories.

vision of "the appropriate nature of science" (in this case, a vision of empirical, limited knowledge) as an influence on the selection of particular strategies for defining and defending that vision. What was at stake in the Darwinian controversy, Campbell (1975) seems to suggest, was not simply the public's acceptance of a particular theory of natural history but also the veracity of a particular understanding of science. In this way, the issue of demarcation is implied along the conceptual fringes of Campbell's accounts of Darwinism.

This indirect connection between the Darwinian debates and demarcation issues is perhaps better exemplified elsewhere (1974, 1989). In his first essay, Campbell (1974) argues that the profound impact that Darwinian ideas exerted on the scientific and public communities of their age serves as prima facie evidence for the suitability of an investigation of those ideas in a different intellectual milieu, that is, the present. Campbell situates his analysis of the *The Origin of Species* not in the rhetorical situation of the mid-nineteenth century, but in the current day. He argues that the contemporary reader is granted a particular view of the relationship of humankind and nature (as mediated by science) which is particularly compelling in its implications for crafting a constructive approach to the contemporary ecological crisis.

Campbell's analysis suggests a meaningful "polysemy" of science (see Fiske and Hartley 1987; Condit 1989). Science, as operationalized in the mid-nineteenth century, became invested with fundamentally different "meanings" as the broader natural and historical contexts in which it became embedded evolved. Whereas "science" may once have been viewed as the intellectual warrant for the plundering of natural resources, one hopes that it may now become (re)constructed as one of the best hopes for saving what remains of those resources. Science, as a cultural discourse, can be read in many, even contradictory, ways. In this sense, the image of science, as defined by particular demarcation strategies, is a product of both historical evolution and rhetorical negotiation.

The final part of Campbell's account of Darwin that I wish to bring to bear on the current study involves the alternative (re)constructions of Darwinism which emerged from the prevailing intellectual trends of the mid-nineteenth century. Before the publication of Darwin's *Origin* in 1859, "belief in the divinely ordered harmony of nature and even in the special creation of life forms were tenets not merely of sectarian religion but of science itself" (Campbell 1986, 351). The fact that not more than two decades later the evolutionary hypothesis had become, for the most part, tacit social knowledge thus represents an intriguing mystery. The solution to that mystery, according to Campbell, is to be found in the boundaries of the new Darwinian science which were constructed inclusively vis-à-vis the prevailing, albeit misleading, caricature of "Baconian science" (1986, esp. 352–66). This so-called science was grounded ostensibly in commonsense empiricist realism, a strict mind/matter dualism, and rigid adherence to the inductive method. "Hypothetical" Darwinism, of course, could have represented a fundamental challenge to these accepted dicta, or demarcation criteria, of science. As such, Darwin's rhetorical exigence was a formidable one, for "not simply the misunderstandings of science on the part of popular religion, but the moral and social aims of science itself were at issue in the debate over the radical alternative to the creation story presented by Darwin's *Origin*" (Campbell 1986, 351). Clearly, competing visions of science and, more particularly, its relationship to and/or primacy over other forms of culture were being contested.

Viewed in this sense, the crisis represented an exemplary case study in the practical management of demarcation. It was through the masterful conduct of such demarcation strategies that Darwin's *Origin* became so rapidly accepted as a cornerstone of the scientific and social communities. Darwin did not construct his science as antithetical to the prevailing cultural grammar; rather, "deference to Baconianism is one of the most marked features of Darwin's text" (Campbell 1986, 358). Darwinian science, then, in this particular context, was constructed in such a way as to resonate with the most profound commitments of the prevailing natural theology, and a potentially revolutionary scientific development was rhetorically negotiated into the intellectual currency of the day.

This episode implicitly anticipates the practical importance of a rhetorical perspective on demarcation. While not all demarcation episodes require such seemingly conscious intention, it is crucial to observe that it was the nature of the demarcations advocated that influenced the reception of the Darwinian position.[4] Had the demarcation practices constructed a vision of the appropriate nature of science as diametrically and inexorably opposed to prevailing visions, a radically different historical outcome could well have resulted. This suggests the historical variability of the nature of science. To sanction static and ahistorical no-

4. The question of intentionality goes to the heart of Gaonkar's critique of Campbell's work. I would say simply that it seems pointless to deny the constraint potentially exerted by what individual rhetors appear to have intended a particular discourse to accomplish, even conceding the ultimate indeterminacy of our historical reconstructions of those intentions. Of course, as the ecosystem extends to implicate a wider range of institutional practices, the constraint exerted by, and hence the relevance of, intention decreases. In this way, I think Prelli's (1993) claim that Gaonkar's critique assumes artificial polarities is convincing.

tions of science is to ignore the lessons of Campbell's work. A constructively rhetorical understanding of the demarcation of science and other forms of cultural practice appears as a powerful explanatory vehicle in the context of scientific controversy.

One cautionary note regarding my application of Campbell's essay is in order. Campbell does not cast his argument in relation to the question of demarcation. His perspective is drawn primarily from models of conceptual change propounded by Kuhn (1962, 1977) and Toulmin (1972). As a result, he is led to conclude that "if the analysis here is sound, what it argues for is continuity in cultural forms enduring across change in proprietorship of those forms" (Campbell 1986, 369). While it is unclear precisely what is meant by "cultural forms" and their "proprietorship," Campbell's claim seems to imply that science itself has not changed; only variously incomplete (or inaccurate) ways of talking about it have. I recognize and sympathize with the theoretical caution that presumably lies behind this statement. I would suggest, however, that "science" itself, as a reflection of its rhetorically demarcated boundaries, did change. To claim otherwise, it seems to me, is once again arbitrarily to posit a pristine and discrete entity of science which is relatively immune from the vagaries of rhetoric.

A series of essays by Gross (1988, 1989; see also 1984) has a great deal of potential for moving in just such theoretical and critical directions. These essays tend to argue, in examining different controversies in the history of science, that the power of rhetorical discourse extends to the "invention of norms" in contemporaneous understandings of science.

Gross takes up the issue of the rhetorical establishment of a particular convention as a "norm" of the scientific community. In examining the heated controversy between Newton and Leibniz over priority in the invention of calculus, for example, Gross tells the tale of "the rhetorical invention of scientific invention" and suggests that "the story of the emergence and transformation of a social norm is unequivocally a tale of persuasion" (1989, 90). Gross suggests that the Royal Society was founded upon a presumption of cooperation, rather than competition, in the pursuit and production of knowledge. While mild conflicts had erupted and had been settled quietly before, the Leibniz/Newton controversy led to a spirit of competitiveness that seriously threatened "to undermine the purpose of the enterprise" (90). As broader cultural factors began to impinge upon the Society, it chose dated journal publication as the hallmark of priority. Gross argues that this turn of events, involving scientific, popular, economic, and imperialist motives, was fundamentally an exercise in the rhetorical negotiation of a social norm for science.

It is important to note that the priority conflict was not simply a question of how to ascertain who "did it first." The controversy itself implicated and was empowered by opposing perspectives on the true nature of science and scientific method. As Newton argued before the Royal Society, "The one proceeds upon the Evidence arising from Experiments and Phaenomena. . . . the other is taken up with Hypotheses, and propounds them, not by Experiments, but to be believed without Examination" (Newton 1980, 314). These competing models of scientific practice, coupled with nascent xenophobia and economic motivations, suggest that the construction of the social norm of priority was not an objective reflection of self-evidently appropriate scientific practice. It was rather quite clearly a matter of the rhetorical management of competing interests and perspectives within a particular historical context.

Again, a note of caution is warranted, because the relevance of this study for a rhetorical account of demarcation is, at best, indirect. Gross (1989) grounds his analysis of this scientific controversy in Merton's (1973) concept of sociological norms. As I have previously argued, in many ways Merton unreflectively posits such norms as empirically constitutive of the scientific community (Mulkay 1975). Gross concedes that the relationship between the sociological and rhetorical dimensions of the normative structure of science demands greater specification (see 1989, 100-104). A rhetorical study of demarcation holds considerable promise for providing that specification, as it explores the discursive negotiation of complex cognitive, social, technical, and professional norms.

A second case study (Gross 1988) seems to articulate even more self-consciously a constitutively rhetorical conception of science. The study takes as its object of investigation the controversy over Newtonian optics. In 1672 Newton posited a conception of optics which contrasted with the prevailing Cartesian paradigm. In Newton's formulation, white light was revealed as a compound of all dimensions of the spectrum. More important, however, "the ground Newton offered for certainty in optics differed radically from his predecessors. . . . By giving epistemological priority to experiment over rational intuition, Newton overturned a central presupposition of traditional science" (Gross 1988, 2). In that light, the nature of the disagreement between the two positions implicates demarcation. Science, Newton argued, is most appropriately founded upon the epistemological presumptions of experimentation. As such, Newton's discourse represents an exercise in the rhetorical constitution of science via methodological dicta.

The historical record, of course, shows that Newton's initial offering, to borrow Hume's phrase, "fell stillborn from the presses" (Hume 1955, 3), most likely as a consequence of its radical departure from prevailing scientific presumptions. In 1704 Newton published his *Optiks,* in which he downplayed his break with traditional conceptions of science in favor of an "essential continuity." "In his final masterpiece," Gross writes, "Newton transformed optics, and experimental science, by allowing his fellow physicists to believe that an adherence to the new did not entail a fundamental rejection of the old" (1988, 2).

From this careful case study, Gross concludes that "the triumph of the *Optiks* is wholly rhetorical because science is rhetorically constituted, a network of persuasive structures, patterns that extend upward through style and arrangement to invention itself, to science itself" (1988, 13). In this sense, Gross offers a conclusively rhetorical account of scientific practice, even the nature of science itself.

Because such a notion is likely to strike some as far too radical, I should be clear on what I take to be its most important entailments. Gross's assertion is not tantamount to saying that everything is rhetorical. To argue that science is constitutively rhetorical is not to deny the brute fact of empirical regularities in the natural world, for instance. I do not suggest that scientists make it up as they go. Certain recurrent patterns in the natural world, however, do not constitute science; they become science only via processes of interpretation, hence reconstruction. Their application in science is but a by-product of their incorporation by a given theoretical proposition. As Gross puts it, "Science is no open sesame to this real world" (1988, 15).

In addition to serving as the clearest argument for a constitutively rhetorical nature of science, Gross's 1988 study also highlights the contextual variability of demarcation practices. Recall that Newton's initial (and more explicit) demarcation met with little scientific or popular support. This, it seems, can be attributed to the uncomfortable "fit" between the newly defined boundaries of science and the prevailing social and intellectual interests which were implicated and fundamentally challenged by those new boundaries. (On the influence of communal interests on scientific argument, see, e.g., Bazerman 1984; Shapin 1979; and Yearley 1982.) Those interests appeared doomed by the authority of Newton's first essay on optics, leading him to reframe his (re)construction of science so as to allow a rhetorical fit between his truly innovative views and prevailing scientific presumptions. Accordingly, we can argue that science, as a series of cultural practices, cannot and should not be viewed as conceptually and functionally immune from broader social and cultural practices, each of which might be configured within a particular enactment of the ecosystem.[5]

This need not be a matter of "anything goes" (Feyerabend 1975a). As Gross puts it, "The rationality of science consists in the dialectic among its legitimate reconstructions, each the surrogate for the informed assent of an interpretive community; analogously, the objectivity of science is constituted by the configuration of these reconstructions" (1990a, 52). What a rhetorical perspective on demarcation promises is a close analysis of the ways in which those boundaries are defined, argued for, modified, and ultimately maintained or discarded in ongoing practice.

Why a Rhetoric of Science?

As Burkeians (Burke 1962), Freudians, and even method actors can attest, understanding motivation is a difficult task. This is perhaps even more the case when considering the motivations behind particular traditions of scholarship, because the old saw about "the humane search for new knowledge" is as vacuous as it is glorific. So my intent here is not to explain the motivations behind individual research programs. Rather, it is to group general lines of research into two broad categories, constructed on the basis of what I perceive their primary critical focus to be. The first motivational category deals generally with philosophical issues, such as the epistemic standing of rhetoric, science, and the rhetoric of science. The second category is defined by what seems a more avowedly political orientation, for example, to reclaim the "public sphere" or to provide foundations for a critical interrogation of modern techno-scientific culture.[6]

The view of "the rhetoric of science" as anything more than an oxymoron is of relatively recent vintage. For centuries, the privileged (and mystified) epistemic status of science (however defined) was considered sacrosanct. Rhetorical studies, traditionally concerned with *doxa*, the domain of the contingent, were considered inappropriate methods of investigation for discourse thought to fall in the domain of certainty.

5. On this issue, for example, Reeves (1990) has argued convincingly that the first three medical reports on AIDS relied on prevailing (and prejudicial) attitudes toward male homosexual behavior in drawing etiological conclusions.

6. Of course, I recognize that these categories are imperfect. Clearly, philosophy can be (even must be) political, while practical politics is premised on philosophical foundations (of ambiguous ontological status). As Lentricchia observed, "All intellection . . . is a form of political action" (1983, 113). My primary goal in this book is not especially critical or emancipatory, although there are implications within it for both sorts of work.

As Melia points out, however, "The publication in 1962 of Thomas Kuhn's now familiar *The Structure of Scientific Revolutions* changed that" (1984, 303). The aura of certainty which surrounded the proclamations of science had begun to fade, inspiring a fundamental reconceptualization of the relationship of rhetoric, philosophy, and science. The boundaries between rhetoric and philosophy are, of course, a matter of considerable historical controversy. In what follows, I make no pretense at settling that dispute but try to draw out the implications of these philosophical rhetorics of science for a rhetorical perspective on demarcation.

While there is great disparity among the stories described here, a fundamental commitment appears to unify them. The intellectual context for most recent discussions of the rhetoric of science emerges from a postpositivistic *weltanschauung*. I will use the term here to refer to a conceptualization of science as operating in and interpreting the world from sets of interpretive frameworks. Although weltanschauungen theorists differ on particular issues, all challenge certain fundamental assumptions of the received view of positivist science, including the uniformity of nature, the distinction between fact and theory, and the objectivity of science. Within communication studies, Delia has argued that from this perspective, "it is . . . within the researcher's particular theoretic stance and a shared matrix of disciplinary commitments, values, and research exemplars that meaning is given to observations taken in specific investigations" (1977, 68). Orr succinctly describes the implications of this perspective for a rhetorical view of science: "Rhetoric as symbolic advocacy is a constituent element in the social construction of reality; even a scientific community's version of reality depends upon rhetoric" (1978, 263). The centrality of rhetoric to this view is closely related to the primacy accorded to the constructive functions of human language. Suppe argues that "science is done from within a conceptual perspective which . . . provides a way of thinking about a class of phenomena . . . [and] is intimately tied to one's language which conceptually shapes the way one examines the world" (1977, 126). Accordingly, our grasp of realities, even scientific ones, is inextricably linked to the discourses in which those realities are represented. In finest Protagorean fashion, Gross concluded that "if man is the measure of all things, his perceptions are criterial: it is he who determines what exists and what does not" (1991, 284).

Working from a similar stance, though without Gross's ontology, Weimer (1977, 1979, 1980) illustrates philosophically inclined rhetorics of science. Implicating the relationship of rhetoric and science by rejecting the positivist view of the inexorably rational and progressive nature of scientific knowledge, he primarily targets what he labels "justificational" and "neo-justificational" accounts of science, which identify scientific knowledge with formally logical proof and unreflective epistemological authority. In their place, Weimer posits a "non-justificational" rhetoric (see esp. 1977). Such a rhetoric entails that the standard for knowledge claims is that of "warranted assertion" (Weimer 1979, 41). Hence, science is not primarily a matter of proof as it exists in formal logic. Instead, it should be conceptualized as a "rhetorical transaction in the classic Greek sense of the term: it is a non-justificational, argumentative, and persuasive art form whose reasoning is adjunctive and whose tuition and communication are injunctive" (Weimer 1977, 14).

A related element of Weimer's nonjustificational rhetoric of science involves the education of neophyte scientists. He notes that "the argumentative mode of discourse requires injunction rather than description." The discursive structure of research articles and the educational process constrain novices to "do this and . . . experience the world correctly" (Weimer 1977, 12). Presumably, then, the educational and discursive processes of the scientific community actively construct a normative conception of science which scientists "in the making" are compelled institutionally to elaborate.

At base, this is a thoroughly rhetorical account of science and scientific knowledge. Indeed, Weimer argued that "with the abandonment of justificationism, it becomes 'self-evident' that the logic of science is actually rhetoric" (1979, 78).[7] An even more radical (and heuristically insightful) perspective is implied in his contention that "the argumentative mode of discourse . . . is literally the essence of science" (Weimer 1977, 13).

Although Weimer never explicitly addresses the demarcation of science, implicit themes in his work bear on demarcation questions. Initially, Weimer's general perspective on the "argumentative essence" of science actively militates against any objective set of demarcation criteria. To extend (my interpretation of) Weimer's position, science is an argumentative construction. While such criteria as theoretical simplicity and falsifiability might well be (nearly) uniformly accepted as constitutive of science, their power lies not in their inherent "scientificness," but rather in their argumentative force. That argumentative force derives its strength not from nature but from the everyday productive and evaluative contingencies of the scientific community.

Additionally, Weimer's discussion of the related processes of socialization and education operative in fields traditionally deemed scientific is an implicit indication of the demarcation practices of scientists.

7. The reference to "rhetoric," unfortunately, is left unspecified. It is this analytic blind spot that, for example, Prelli's (1989b) discussion of "topical logic" illuminates.

If we take seriously Weimer's claim that scientific education is fundamentally injunctive rather than descriptive, then we can note the actual construction of the normative boundaries of appropriate conduct within science. In a very real sense, those normative boundaries serve to define science for those who claim to do science. Recall Mulkay and Gilbert's (1982b) discussion of "accounting for error." They suggested that traditional approaches to demarcation assume synonymy between normative and ontological commitments; that is, appropriate practices are thought to follow from and are evaluated according to their "fit" with nature. Appropriate practices produce scientific knowledge, while inappropriate ones constitute the locus of error and must be explained away, as nonscience. Understood in this way, the rhetorical construction of standards for the appropriate nature of scientific practice involves the implicit management of demarcation.

Despite their contributions, the epistemological works of Weimer and others (e.g., Kelso 1980) become bit players in the rhetoric of science versus epistemology debate when Gross (1990a, 1990b, 1991) enters the fray. His is perhaps the most fully developed and vigorously defended account of "radical [rhetorical] relativism" (Gross 1991, 284). As Gross tells it, this account is full of sound and fury and potentially signifies a great deal: "By means of the rhetorical analysis of the hard sciences . . . rhetoric of inquiry inserts itself into the inner sanctum of epistemological and ontological privilege" (285).

His sophistic epistemology calls into question the purportedly conclusive claims of rhetoricians espousing even a "moderate" realism (Gross 1991, 285). Beginning with claims of empirical regularities which should militate against radical relativism, Gross reconsiders the (dis) continuities between ancient Babylonian, ancient Greek, and Newtonian astronomies. While all were able to adduce roughly reliable predictions, they began from fundamentally opposed assumptions. From this, he concludes that "prediction, however accurate, tells us nothing about truth" (289).

Similarly, Gross rejects even the common claim that surely we must admit the ontological reality of some phenomena, such as electrons, because observation of such "unobservables" requires instrumental interventions such as an electron gun and the subsequent interpretation of the data offered up by those interventions. This characteristic, Gross insists, reinserts rhetorical processes into the very heart of ontology: "For science, it is not the tracks [of electrons] that matter, it is their selection as a target of explanation, and their interpretation within some current theory." Those processes are "well within the boundaries of . . . rhetorical analysis" (1991, 293).

Having slain two traditional dragons in his epistemo-rhetorical quest, Gross bumps up against the most fearsome beast yet—"recalcitrance" (McGuire and Melia 1991), or the general view that humans are constrained by nature. (For example, we typically don't deny the reality of brick walls and continue to pound our heads against them to illustrate our commitment to that belief.) Gross's radical relativism, however, avoids mortal combat with recalcitrance by finding sanctuary in the mundane practices of science (1991, 296). Drawing on sociological studies of experimental practice "at its creative edge," Gross retorts that "the confirmation of theories of great generality . . . are little endangered by experience" (297). In short, much of what we are perfectly comfortable calling science is not ontologically dependent on the recalcitrances of nature.

This, then, brings Gross's rhetoric of science full circle. Effectively rejecting continuity and ontology as sufficient to close down scientific debate, Gross returns to his sophistic foundations. Surveying the pervasive disagreement about the "texts" of science, Gross raises the rhetorical banner: "Each must be interpreted. When interpretations differ, there is but one means of settlement: persuasion, the art of rhetoric" (1991, 284).

The rhetoric of science gives primacy to the epistemic authority of rhetoric, rather than to the ontological constraints of the natural world that traditional epistemologies privilege. Of course, such questions do not bear directly on the definition of science. Or do they? Certainly the aims are not identical. Nonetheless, Weimer's discussion of the essence of science as inhering in rhetorical constructions of appropriateness suggests that what we come to take as science is a product of contextual rhetorical negotiation. Similarly, Kelso's (1980) call for explication of the bases of epistemological privilege foregrounds a concern with the boundaries of science. Most notably, Gross's insistence on thoroughgoing rhetoricity entails a commitment to the ultimate indeterminacy of the nature of those practices which we label honorifically as "scientific." While I am neither concerned nor qualified to adjudicate the ontological status of quarks or fusion, the lessons of such a radical rhetoric of science extend comfortably to questions of science's own communal definitions.

In various ways and to varying degrees, many have questioned the insights of the weltanschauungen perspective in general and the rhetoric of science's application of it in particular. In his critique of the rhetoric of science literature, Bokeno, for example, is more concerned to reject the weltanschauungen orientation, which is said to undergird contemporary rhetoric of science. Situating himself in what he calls postweltanschauungen philosophy of science (see, e.g., Bhaskar 1978; Newton-Smith 1981; Russman 1987), he argues that "the price for . . . success in recovering the hermeneutic dimension of science has been a relativism which provides a troublesome account of the nature and growth of scientific knowledge" (Bokeno 1987, 295). At base, he argues that the weltanschauungen program "hinges on the idea of meaning," entailing that "meanings provide the only available interpretive context for members of that [scientific] community to perceive, describe, and explain phenomena" (296). Bokeno extends his attack to what he calls "conceptual relativism: the concepts available to a particular community will constitute *what is seen* in reality" (297). Thus he contends that such a position cannot supply an acceptable account of the nature of scientific knowledge and the progress of science. If, he argues, scientists with different interpretive frames actually see different worlds, then the obvious success or progress of science cannot be explained rationally.

Bokeno ends his argument precisely where McGuire and Melia (1989, 1991) begin theirs. Motivated by a skepticism of the "too easy assumption that scientific texts are as susceptible to rhetorical analysis as are texts in other disciplines" (1989, 87), they steer their rhetorical ship by the twin stars of recalcitrance and minimal realism. The moral of their story is admonitory, warning against the intellectual dangers of Gross's "rampant rhetoricism" (McGuire and Melia 1991, 301).

While eschewing simplistic tests of empiricism, they insist that human actors, whatever their rhetorical skill, are inescapably constrained by the empirical world. "In our view," they say, "our beliefs and practices do interact and do collide with a nonhuman world, but in virtue of links that are causal and nonintentional and nonrepresentational." These interactions and collisions lead them to maintain that "we are pressed into shape by recalcitrances that are not of our free choosing" (1991, 308). The story of science, then, whatever its rhetoricity, must still be told from an empirical platform: "Science not only encodes relationships linguistically. . . . it also predictively and creatively interferes and intervenes with nature" (1989, 99).

From this, McGuire and Melia reason that our rhetorical studies of science must allow for the discovery, rather than simply the construction, of facts. As they describe it, "The *discovery* of such facts comprises so large a part of the quotidian practice of science that any rhetorical account that neglects them risks rendering itself irrelevant" (1991, 304). That irrelevance would be purchased at the cost of ignoring what they take to be an undeniable fact—that science can, does, and will continue to explain the natural world with relatively frequent success.

They contend that relativists such as Gross assume wrongly that the explanatory success of science is a marker of truth. For minimal realists that success should be understood in a less grandiose way, as providing the grounds, "in virtue of the success over time of a particular piece of science, for supposing that the kinds of entities it posits *exist*." Existence, then, is construed as a prerequisite for a meaningful rhetoric of science; truth is not. As they concluded, "We should never infer from the fact that our practices are necessary for access to scientific entities, to the conclusion that these entities are entirely *constituted* by our practices" (McGuire and Melia 1991, 310).

This line of argument seems flawed on two dimensions. First, Bokeno and, to a lesser degree, McGuire and Melia assume that the constraints exerted by material reality is rendered trivial by weltanschauungen philosophy of science and by rhetorical studies of science issuing from it. Such need not be the case. As I noted above, one need simply insist that such exertions are not experienced directly, discretely, unproblematically, or nonsymbolically. To suggest that humans are interpretive creatures is not to say that reality does not impinge upon them at all, only that it does so in symbolically mediated fashion. The negotiation of the particular structures we come to take as scientific involves demarcation and is intrinsically rhetorical.

While I am not at all convinced that engaging, less yet settling, this epistemological issue is a prerequisite to doing rhetorical analyses of scientific discourse, some position is required. If I describe scientific culture as rhetorical, there is a corresponding temptation to assume that all scientific practices and products must emerge from a "radical rhetoric of science," one that assumes all is rhetorical, without remainder (McGuire and Melia 1991). I think we must resist this temptation. The conception of science developed here (as a series of practices rhetorically defined in alternative ways) certainly expands the ranges of discourses considered "scientific," but it does not entail that all of the practices within those boundaries are reducible to rhetoric. Technical practices such as constructing bubble chambers or preparing brain extracts don't strike me as rhetorical, or even as remotely discursive. Certainly, the network of propositions which both motivate those behaviors and subsequently issue from them is rhetorical. Nonetheless, the acts themselves appear to be both nontrivial and nonrhetorical. Or, to beg the definitional question, they can be studied more productively from perspectives other than those adopted by most rhetorical critics.[8]

8. For example, recent work in cognitive studies of science, or the psychology of experimentation (Fuller 1992) might hold promise for explicating the intrinsic character of these practices. The ethnographic work of Lynch, Latour, Woolgar, and others seems similarly equipped for such commentary.

I think we set for ourselves a unnecessary and unproductive burden if we assume that the telos of the rhetoric of science must be the explanation of the real, albeit rhetorical, nature of science and scientific facts. This position—and I plead guilty to having committed the infraction myself—entails that we must somehow hew to traditional ways of mapping the intellectual geography. I am more inclined to believe that the shifting natures of science, rhetoric, and knowledge militate against such an approach, suggesting the utility of oppositional, at times counterintuitive, reading strategies. I hold an agnostic position on whether the true nature of science (or any cultural practice) will ever be fully known. My point is that, for scholars of rhetoric, a more promising focus lies with the discursive practices of science.

This is not to say that the analysis of scientific rhetoric is a mere adjunct to the real business of science (or, less yet, of "science studies").[9] The wealth of critical studies leads me to insist that its most important aspects are inescapably rhetorical. Nonetheless, extreme formulations of the "all science is wholly rhetorical" position seem to define rhetoric so broadly (or science so narrowly) that both lose much of their meaning and, worse yet, set as a goal a critical burden impossible to meet, even for the best rhetoricians, sociologists, or philosophers. I'm not sure that rhetorical hubris is necessarily superior to its philosophical or sociological incarnations.[10]

Of course, I need not and cannot hold that science is just like every other form of life. If it were, the question of demarcation need not come up at all; there would be nothing from which to demarcate science in the first place. Indeed, the sheer epistemic privilege typically granted to science suggests that it is a social construction that is taken to be qualitatively different from other social constructions. As a "practical matter" (Gieryn 1983a, 781), science *is* demarcated from other forms of

9. This seems to be the general thrust of Myers' criticism of the rhetorics of science and inquiry. See his 1990 review of Prelli and his 1989 study of power.

10. Most emphatically, this does not entail that the intellectual burdens for rhetoricians of science are any lighter. Gross is absolutely correct when he insists that "there is nothing wrong with rhetoricians doing philosophy, so long as they get it right" (1990b, 305). It simply means that we should highlight our guiding concern with the discursive practices of science as legitimate foci of study in their own right, without necessarily claiming that such foci depend upon or supersede others claimed in and by other disciplines. At this point, I echo McGuire and Melia's claim (absent the minimal realism) that "rhetoric . . . should, by pressing the claim of proportion rather than limits, resist the very idea of disciplinary hegemony" (1991, 316). Indeed, Pickering has argued that "the study of [scientific] practice works to undermine traditional disciplinary reductions" (1992, 7). Perhaps rhetoric will turn out to provide the story to end all stories. However, we shouldn't assume that it will.

life and knowledge. After all, the National Science Foundation typically does not provide grants to sculptors, philosophers, or rhetorical critics. The crucial question then is how, and on what symbolic foundations, is this demarcation practically accomplished? Answering this question does not require the reification of science as a discrete and unproblematic lifeline to the natural world. I need not entertain the possibility of universal standards for science to get on with the critical analysis of the rhetorical negotiation of the social boundaries which have come to delimit science.

The tangled connections of rhetoric, science, philosophy, and knowing have proved a critical concern for the rhetoric of science. Other research has extended this concern to politics in the making. Viewed in this way, "rhetoric of inquiry is a way of conversing about intellectual conversation—and improving its quality" (Nelson, Megill, and McCloskey 1987, 4).[11]

The "Spheres" Debate and the Rhetoric of Science

The general exigence to which these studies respond was described in one of the earliest forays into the field, Wander's "The Rhetoric of Science" (1976). A keen observer of the "ideological turn" in rhetorical studies, Wander is particularly concerned with the sociopolitical impact of scientific forms of knowledge. He notes that "reality and the rules by which it is validated are everywhere cloaked with an air of mystification. . . . On every great issue . . . science, scientists, and scientific thinking shape the debate" (1976, 226). Science, then, is articulated as a validated representation of nature and the methods for that validation.

This particular element of Wander's analysis is relevant to the current project insofar as it suggests that the nature of science is (or at least can be) subject to (re)construction. Of course, Wander's concern is with manipulative constructions. We need not be limited by a necessarily "evil" view of science's epistemic privilege, however. At present, I wish to call attention to its implications for a dynamic conception of science.

11. Of course, the setting in which that "improvement" is enacted is quite varied. For rhetoricians with backgrounds in English and/or composition, the practical import is with classroom writing instruction. Bazerman (1987), for example, understands his work as implicating the very practical problem of teaching technical writing to undergraduates. On the pedagogical applications of the rhetoric of science, also consult Berkenkotter, Huckin, and Ackerman 1991; and Schwegler and Shamoon 1991.

Wander's analysis does not implicitly sanction a static conception of science. Rather, it situates the socioepistemic influence of science in variably constructed demarcations of science. While Wander's most direct interest is with the influence of science on public deliberation, he does make some observations regarding the discursive strategies utilized by scientists themselves. In his discussion of the "rhetoric of theory," for example, Wander seems to imply that even the constitutive theoretical assumptions of traditional science are rhetorical constructions which function as "telescoped rationalizations" (1976, 232).

It is important to note that Wander does not posit a continuous and objective domain of science. His essay suggests a recognition of the political implications of particular constructions of science and scientific knowledge. While much of its insight remains tacit, Wander's essay should be considered valuable insofar as it establishes a theoretical and political background for a meaningfully rhetorical conception of science and its various demarcations. The way in which we determine what may be legitimately taken as science is a consequential matter and one warranting sustained critical attention.

The previous section on the contextuality of scientific discourse has already considered, at least indirectly, the potential impact of the sociopolitical milieu in which scientific practices are embedded. My concern is this section will be to comment briefly on the body of literature which takes as its focus the distinctions and relationships between scientific (usually characterized as technical) and public spheres of discourse and/or reasoning. My goal will not be to offer a definitive contribution to the ongoing debate over "spheres" of discourse, but rather to incorporate its nuanced motivations into a broader understanding of rhetorical demarcation practices.

Goodnight and Farrell can be seen as the exemplars of the burgeoning "spheres" literature (Goodnight 1982, 1988; Farrell 1976; Farrell and Goodnight 1981). While there are conceptual differences between the various positions (e.g., T. Peters 1989), Goodnight has most clearly cast the "spheres" debate in its current terms. He suggests that a sphere denotes branches of activity—the grounds upon which arguments are built and the authorities to which arguers appeal. He is concerned fundamentally to recover and rehabilitate the public sphere as the site at which "citizens test and create social knowledge in order to uncover, assess and resolve problems" (1982, 214). Standing in opposition to the attainment of this goal, he maintains, is the encroachment of the technical sphere.

Differences among these three spheres—personal, technical, and public—may be plausibly illustrated if we consider the differences among "the standards for arguments between friends versus those for judgments of academic arguments versus those for judging political disputes" (Goodnight 1982, 216). In the technical sphere, Goodnight maintains that "the concrete particularity of the original dispute is lost" and the dispute is moved to a realm of professional insulation. There is the assumption, however, of "a special kind of understanding among members of a professional community" (219). The public sphere, on the other hand, is seen as transcending the private and technical spheres and becomes operative when "neither informal disagreement nor theoretical contention is sufficient to contain the arguments involved" (223).

Contributing an agonistic dimension to this story is its typically oppositional pairing of the public and technical spheres of argument. Goodnight argues that "if the public sphere is to be revitalized, then those practices which replace deliberative rhetoric by substituting alternative modes of invention and restricting subject matter need to be uncovered and critiqued" (1982, 227). Farrell and Goodnight, reflecting on the Three Mile Island debacle, observe that "the limits of technical communicative discourse are severe, recurrent, and irreparable" (1981, 271). When public advocates do manage to make meager contributions, some have suggested that they "bias expert decision making" (Rowland 1986, 139) or promote discursive strategies which "seriously hamper intelligent discussion of crucial public issues" (Bantz 1981, 87). Other commentators, however, have held out more optimistic readings of the vibrancy of public deliberation in the technological age. Balthrop argued that we "turn to the public sphere to define the limits of the technical and private," so that we can "form not a perfect union, but a more perfect union" (1989, 25). Similarly, Waddell's (1990) study of ethos in the Cambridge recombinant DNA controversy illustrates a fruitful symbiosis of public and technical deliberations.

While there is much of value to this "tale of three spheres," it should be noted that the "spheres" literature is not conceived as a contribution to the demarcation debate. Clearly, we cannot take the suggested criteria for the technical spheres, such as relatively high degrees of required preparation, clarity of argument, specifiable rules of argument, and so on, as constitutive of science. Indeed, the technical sphere is taken as any specialized discourse community in which mastery of a particular body of knowledge is presupposed. In that sense, an argument regarding the proper lineup for the 1996 Olympic Dream Team might be considered technical in nature, especially if the argument took place among members of the team or other basketball insiders. The dispute would not, however, typically be considered scientific, or even especially interesting. This is not to demean the insights of Goodnight and others. It is merely to concede that we are telling different stories, but ones that have much to say to each other.

In that spirit, several implications of the literature merit discussion here. First, it should be noted that the oppositional perspective detailed above and Goodnight's (1988) broader concern with the recovery of the public sphere tend easily to lead to an unjustifiably broad skepticism of the technical sphere, which can cause us to lose sight of the legitimate contributions made by technical ways of knowing (Willard 1985, 442–43). Such a skepticism suggests that technical discourses (of which scientific discourses might be considered the exemplar) are to be studied only in order to guard against them. This seems to me unduly pessimistic. At the very least, it would seem that an investigation of the processes wherein working definitions of science empower certain technical discourses would facilitate the "uncovering" and "critique" of those discourses that Goodnight (1982) deemed necessary. A rhetorical study of demarcation can lead to a more reflective and, as a consequence, a more humane understanding of social epistemic authority.

I suspect that some might sense more than a whiff of giddy naïveté here, supposing that somehow this enacts a sort of rhetorician of Sunnybrook Farm mentality. McGee and Lyne (1987), for example, have suggested that the public and technical spheres should remain in "dialectical tension" (389), and the likelihood exists that public rhetorics will "dangerously oversimplify" (400) the issues confronting us. I think the cogency of their claim stems from current failures to articulate clearly the grounds for demarcating the two. If we adopt a rhetorical conception of demarcation, we reinsert the human element into a heretofore mystified technical sphere. As Fuller has argued, "The failure to scrutinize scientific practice reinforces the myth of . . . expertise" (1992, 422). The implications of this issue are important. They suggest that the distinctions between the spheres are more a matter of relative emphasis on particular pragmatic rules or strategies and not a matter of conceptual distinctions. Accordingly, the problem might simply be that people tend to think of science as an activity sui generis, mainly because our rhetorics of and about science have assumed necessary and systematic differences between it and other cultural practices.

In their work on the rhetorics of punctuated equilibrium (Lyne and Howe 1986) and, more recently, on sociobiology, for example, Howe and Lyne (1992a) map the symbolic transformations that they see as accompanying the circulation of "scientific" claims among and between various spheres of discourse or, in rhetorical terms, different audiences. For example, they claim that a genetic metaphor is used frequently to legitimate sociobiology, as "it goes galloping across a range of different audiences . . . independent of precision and controls" (1992a, 121). They further assert that "genetic terminology used as metaphor leads to misunderstanding of projected expertise, and consequently to miscommunication of evidence and authority" (Howe and Lyne 1992b, 231).

Central to this critique, as I read it, are the tacit boundaries that mark the contexts in which the language of genetics is used appropriately, functioning *via negativa* as the standard against which the "gene talk" of Richard Dawkins and others can be found wanting. This grounds the critique upon a global conception of science, though one enacted here through the transcendence of its constituent communities. As Lyne puts it, "This breaking up of sub-specialties . . . is in many ways made possible because of the supposed possibility of reduction. This is what makes each separate activity part of a single enterprise" (1994, 22).

As I see it, an important limitation of this interpretive vocabulary is that, while rightly problematizing sociobiologists' "gene talk," it fails to recognize (at least explicitly) that "genetics talk" is equally a consequence of the temporary stabilization of a particular ecosystem, equally a locally validated set of professional practices. Deproblematizing these evaluative rhetorics in effect lodges agency for judgment in nature, which, we must assume, speaks clearly and with only minor dialect shifts, to the *real* scientists. Consider, for example, the strategy of delineating "the genetic rhetorics that perfuse texts of prominent sociobiologists, and compar[ing] them with their authorizing principles in genetics" (Howe and Lyne 1994, 016). Remarkably, we are told that gene talk's inadequacies are attested not to rhetors but by the facts, represented here as the necessary products of a globalized science. Viewing those facts as the products of local judgments allows one to conclude (with equal empirical legitimacy) that authorizing and authorized principals have evaluated and delegitimated sociobiological claims. Principles do not close down debates; principals, via judicious and persuasive selection and application of competing principles, do.

Of course, someone might well respond that such heterogeneity is getting a bit carried away with itself, that there are lots of genetics laboratories but they do have a great deal more in common than I suggest. Perhaps. However, the most important thing they share would be access to the empirical data, interpretations and evaluations of which might differ considerably. Indeed, what counts as the relevant probative context is a rhetorical accomplishment, not an inherent property of a given context, less yet of the claims circulating within it. The availability of that empirical data is itself the outcome of stabilized configurations that extend well beyond the antiseptic confines of the lab. As Howe and Lyne have shown, for example, "the pragmatics of funding patterns have conspired with demands of experimental rigor to produce dramatically uneven progress in subdisciplines of genetics." As a result, "the uneven progress within subdisciplines . . . has produced rather different vocabularies of . . . critical terms" (1994, 6). If such is the case, one can conclude that, while members of professional communities are the best evaluators of their technical data, the generation of that data is connected inextricably and nontrivially to the funds available to support certain research programs rather than others.[12]

Hence, it would seem that the empirical foundation upon which Howe and Lyne premise their critique is itself pragmatically contingent on much wider configurations of practices—many of which require judgments from those who have never set foot in a genetics laboratory, belying the technical versus public dichotomizing characteristic of this sort of work.

My point here is not that Howe and Lyne are wrong; on the contrary, a more practical consideration of the empirical basis of their own critique might well allow closer scrutiny of the genetic evidence, warranting perhaps a more rigorous critique of sociobiologists' claims of genetic authority. Similarly, I do not want to suggest that all interpretations of the "facts" are of equal legitimacy. That sort of vicious relativism is as discredited as its positivist alter ego. My point is that to claim that relative legitimacy of a given interpretation is a natural condition of the material to be interpreted or even of the context in which said interpretation occurs vests the privileged few with literal "ownership" of knowledge claims. The rest of us, then, are expected to toil without need of understanding, less yet an opportunity for meaningful critique on these epistemological grounds. This is thus a sort of intellectual feudalism. Such a situation, I think, borders on the scenario against which Leff warned: "Lacking a model that provides entry into the world of civic action, the rhetoric of the human sciences would leave human scientists isolated in the web of their own technical practices" (1987b, 30). On the other hand, if we adopt a reading strategy that characterizes the rhetorical transformations that (dis)empower various ecosystems, we leave

12. An additional instance of this nonreflexive discussion of the basis of empirical critique can be seen in the discussion of the practice of extrapolating from animal to human, hence socially shaped, populations. Howe and Lyne write that "the direct application of molecular genetic discoveries to the behavioral ecology of complex organisms is severely limited. Convenient experimental organisms . . . for inherently *expensive* research allow few intriguing inferences relevant to bird or mammal, much less human social behavior. The financial strictures on molecular genetics of birds and mammals are prohibitive" (1992a, 124). Viewed in this way, the technical practices of genetic ecologists are constrained a priori by material practices of social, not natural, origin.

open the potential for crafting such a model. On this, Lyne and I would agree. As he puts it, "Rather than dismissing the rhetorical forces that combine to make a scientific story persuasive, one needs to take them seriously and engage . . . them in critique" (1994, 24).

Perhaps what is called for is less a reaction against the technical sphere and more a critical sensitivity to the common constitutive influences of rhetorical discourse within and among the spheres. Willard (1985), disputing the "false contrast of values with facts, of science with humane concerns" (442), argues that "methods and structures of discourse do not only use values, they create them" (435). The insistence of Goodnight and others on the reification of the technical sphere actively militates against a meaningful understanding of the truly constructive nature of rhetorical discourse in scientific communities. That artificial reification shields the human dimension of scientific practice. To obscure the rhetorically constructed values of the scientific community in this way reinforces its further reification. As Willard argues, "Values emanate from practice and become sanctified with time. The more they recede into the background, the more taken for granted they become" (1985, 441). This story desanctifies these values by bringing their rhetorical origins into the foreground of critical scrutiny.

Of course, any such attempt at desanctification assumes a working knowledge of how the prior sanctification was itself accomplished. Thus, we are led to the myriad "public rhetorics" of science that are among the most powerful mediators of scientific knowledge for nonexpert audiences (Lyne and Howe 1990; Nelkin 1987; LaFollette 1990). That, however, brings us back to precisely the false public/technical dilemma that we're trying to avoid. While I don't want to plunge into the tangled thickets of science journalism, public understanding of science, and the like, I do want to suggest that the "falsity" of the contrast I have outlined lies less in the practical differences between ways of going on in the world than in the traditional hierarchial arrangement of those differences. Put another way, the fundamental limitation I see with the spheres literature is that it assumes, at least implicitly, a unidirectional process of influence, from the technical to the public, without a clear recognition of the synergistic mutuality of the two domains.

The construction of hierarchy might well be a intrinsic function of language, in any case, this particular hierarchy is certainly of rhetorical construction. There is a large (and growing) literature which purports to detail (whether to praise or decry) the mediated vehicles in and through which the public comes to develop an understanding and/or misunderstanding of science and its alleged difference from social life (e.g., R. Anderson 1970; Basalla 1976; Bastide 1992; Brown and Crable

1973; Culliton 1978; Etzioni and Nunn 1976; Graham 1978; LaFollette 1990; Lessl 1989; Nelkin 1987; Taylor and Condit 1988).

There is certainly much to recommend greater attention to ensuring higher levels of scientific literacy in a technological age (Arons 1983; Michael 1992). This is not to privilege what scientists say as having some sort of epistemological or representational superiority (Trachtman 1981). It simply suggests that the definitions of appropriate scientific practice which are rhetorically negotiated by practitioners quite often form the content of public accounts of those practices, and as a consequence, some foundation for evaluation seems desirable.

It is not at all clear that such a foundation is the likely outcome of the hierarchical public rhetorics of science. Lessl (1989), for example, speaks of the public discourse of scientists as framed in a "priestly voice" that neither encounters nor brooks significant public interference. Indeed, an important function of such public rhetorics is the clarification (construction and reconstruction) for nonexpert audiences of the meaning of science. Lessl argues that "rather than conceiving of public science as the popularization of technical knowledge, we might better conceive of it as the scientization of the public consciousness." The practical effect of such discourse is "to make to people suitable for science" (1989, 195).

Similar themes are articulated in the literature on "accommodating science" (Fahnestock 1986), that is, on the discursive transformations enacted as scientific information is communicated in different contexts (Lyne and Howe 1990; Gieryn and Figert 1990; Bastide 1992). In general, such studies point to rhetorical characteristics such as the exaggeration of certainty (Fahnestock 1986, 280), establishment of credibility (Gieryn and Figert 1990), and strategies of inclusion/exclusion (Bastide 1992) as examples of how science is accommodated to the demands of nonscientific publics. What seems especially significant here is not so much the old saw that rhetoric is constrained by context. Rather, it is that speaking of the accommodation *of science* constructs implicitly a hierarchy in which the epistemic privilege of science is threatened by the routines of popularization, enacting, often in spite of its best efforts, more or less sophisticated processes of "dumbing down." The public, meanwhile, remains solely at the intellectual mercy, so to speak, of technical discourses, be they of professional journalists (Gieryn and Figert 1990) or prominent scientists (Fahnestock 1986).

It is somewhat ironic that, in our efforts to reclaim the legitimacy of the public sphere, or at least to craft new ways of engaging in useful public deliberation, we too often ignore the operative influence of public factors in technical debate. Perhaps in attempting to redeem the polis, we should avoid a nostalgia for a time that never existed in favor of mapping discursive trajectories that trade on the interpenetration of traditionally polarized categories. Scientific dispute is not immune from the characteristics that denote the public and personal spheres of argument. Their latent interconnectedness is illuminated and celebrated by a rhetorical account of demarcation.

Rhetoric and Demarcation

The foregoing suggests that the rhetoric of science must move beyond formulary distinctions between different sets of social practices and that we must begin a search for the rhetorical processes in which those distinctions are articulated and legitimated. The practical negotiation of demarcation criteria by those working in traditionally scientific fields is a crucial preliminary step in the explication and demystification of the bases of epistemic privilege in contemporary culture.

While the rhetoric of science literature has only indirectly considered the topics of the internal and/or external demarcation of science, three recent essays (Lessl 1988; Prelli 1989a; Holmquest 1990) speculate on their own relevance to the demarcation issue. Here I want to consider each in turn.

In his account of the traditional scientific community's response to the "heresy of creationism," Lessl (1988) contends that the "public rhetoric" of traditional science is oriented toward the maintenance of the material benefits (e.g., grant money, epistemic authority, etc.) to which scientists have become accustomed. Insofar as creationism is seen as a threat to certain of those interests, it necessitates a response. By way of response, Lessl argues, the scientific community underwent a process of self-analysis and reification. The processes of self-analysis and reification, Lessl suggests, are related to the boundary concerns of sociologists such as Gieryn (1983a, 1983b; see also Gieryn, Bevins, and Zehr 1985).

While there is much to recommend Lessl's analysis, it has several conceptual shortcomings when viewed through a broader rhetorical perspective on demarcation. Initially, his adoption of the "heresy" metaphor, while inventionally efficient, leads to confusion along demarcation lines. The analytic use of the metaphor ultimately leads Lessl to confuse "solidarity" with demarcation. In his account of how scientists rallied 'round the Darwinian flag, he does not extend his analysis to the entailments of the presumptions for which the flag served as symbol. There is little speculation, for example, about the implications of the rhetorical tribute paid to Sir Karl Popper and the falsification principle as "proof" of the pseudoscientific status of creationism. In short, Popper's authority to define the terms of the debate was taken for granted. The heresy metaphor caused Lessl to focus less on constitutive demarcation issues and more on the presence of "error" in science. Lessl, then, like most authors in this tradition, stops short of endorsing a truly constructive view of the rhetoric of science. His analysis suggests a concern for how scientists talk *about* a scientific community which he (and they) assume to exist in some quasi-objectivist form (see Lessl 1988, esp. 20–27).

A final conceptual shortcoming of Lessl's perspective (at least as it bears on the current project) is that it appears to posit a relatively unilateral conception of demarcation or boundary work practices. While he makes passing reference to the "dialogic" nature of the creationist versus evolutionist controversy, he does not give any coherent analysis of the ways in which the nature of science was negotiated, as opposed to declared. Granted, creationism represented the incipient crisis which led to the use of certain demarcation strategies. While interest and space limitations might have constrained him to focus only on the evolutionist response, this is but one dimension of a rather complex story. The rhetorics of creationism served as the rhetorical backdrop against which various conceptions of science were rhetorically constructed and deconstructed.

In his study of the "rhetorical construction of scientific ethos," Prelli also implicates certain questions relevant to the present study. He takes a Mertonian perspective on the normative structure of science, arguing that "when scientists resort to . . . common themes in discussing, justifying, or evaluating actions, the alleged norms and counter-norms of science serve a *rhetorical* function, regardless of whatever other function they might . . . serve" (1989a, 49). That theoretical scheme is the backdrop for his analysis of the controversy surrounding Patterson and Linden's *The Education of Koko* and the general area of ape language acquisition studies. Prelli considers demarcation criteria to be among the many scientific topoi that scientific rhetors may summon in various rhetorical situations, suggesting that attempts may be made to sharpen or blur the boundaries of science, given the particular interests and goals at stake in a given situation (1989a, 52).

Prelli's study makes a valuable contribution to the conceptual background of a rhetorical demarcation narrative. Nonetheless, there are also important limitations to his project. Initially, and like the majority of writers in the area, he implicitly sanctions a static conception of science. His "topical" format is a prime example of this confusion. He consistently maintains that scientists may draw upon these resources in order to describe the scientific ethos. The implication here is that rhetorical discourse contributes less to the actual construction of that ethos than to the *description* of an ethos whose nature remains "black boxed."

A second example of this confusion can be seen in his contention that boundary disputes are typically conducted before an audience which does not understand the implicitly "true" nature of science (1989a, 52–53). This implies a particular, even objective, conception of science which is, in principle at least, capable of universal (mis)understanding. Lacking this universal understanding, it is argued, scientists are forced to adopt topical strategies which variously describe the "black-boxed" essence of science and scientific practice.

Most recently, Holmquest's study of the rhetorical *strategy* of boundary work combines Gieryn's early treatment of the topic with the work of Goodnight and others on spheres, in order to argue that "the demarcation of science and the state is a constantly evolving process for different risks at different times" (Holmquest 1990, 238). Her case study of the *Tarasoff* precedent, which created a mental health professional's "duty to warn," suggests that the boundary between scientific and public forms of knowledge is variable and contextually mediated. In *Tarasoff*, the relative degree of certainty with which "scientific" predictions of dangerousness could be made was the subject of dispute between mental health professionals concerned with their own liability and the judicial structures charged with "weighting" such predictions in legal proceedings.

Holmquest's study performs an important function. Most notably, it challenges the traditional (Gieryn 1983a), but mistaken, assumption that boundary work is but mere "style." Invoking instead the classical canon of "invention," Holmquest is able to demonstrate that the stories of demarcation are far from verbal flourishes to the real nature of science. As she tells it, demarcation reconfigures taken-for-granted boundaries in the face of pressing exigencies.

I'm not sure, however, that these stories reach the narrative closure they hint at. Taken as parts of a unified story, they produce a blind spot which, I think, can be eliminated within a broader rhetoric of demarcation. At present, all the rhetorics of demarcation are considered "public rhetorics," largely conceding the stability of that analytic category. This is indeed an important part of the story. This perspective, however, does not account for the everyday productive and evaluative activities of scientists as laborers and, significantly, of the interaction between the two perspectives. The concerns of the audiences implicated in those case studies (legal responsibility, public approbation, etc.) would seem to exhaust but a small range of demarcation rhetorics.

Most notably, these demarcation rhetorics appear motivated by the specific goal of establishing the boundaries of science for audiences that do not participate in the daily activities of science. Those who do participate in them, professional scientists, are not likely to concern themselves explicitly with defining what it *means* to do science. One might say, they just do it. To practicing scientists, demarcation criteria and the like fall into the category of unarticulated "tacit knowledge" (Polanyi 1958).

This should not, however, entail that professional scientists do not accomplish demarcation rhetorically. They simply appear to do so indirectly. In rhetorically confronting the myriad technical exigencies of scientific practice, such as error correction, report writing, and the like, they construct and reconstruct what it means to do science. For example, to reject an experimental knowledge claim because it cannot be replicated, either in theory or practice, portrays replicability as a constitutive characteristic of those knowledge claims that pass muster as scientific. Viewed in this way, demarcation is less a matter of strategy than of ongoing evaluative practice.

At base, past rhetorical treatments of the relationship of science (even as exemplar of technical discourse) and rhetoric have been marred by a static conception of the main characters of their investigation, that is, a static conception of science. Rarely have writers in this tradition attempted to come to grips with the fundamental issue of the constitutive nature of the scientific discourse in which they are interested. Too often, there has been an unreflective assumption about what is taken to be scientific either by scientists or the public at large. It is one thing to suggest that public policy has been co-opted by the forces of instrumental science. It is quite another to explicate the rhetorical constituents of that science that enabled processes of co-optation.

Of course, it might be responded that the typical approach to the classification of discourses of science, for instance, reliance on archetypes such as Darwin and Newton, is suitable insofar as the selected scientists are, almost unanimously, agreed to be scientists. I must admit that such an approach does have presumptive validity and cognitive economy to recommend it. Nonetheless, we should beware endorsing such a strategy without a more critical analysis of its interpretive entailments. First, it presumes a rather ahistorical conception of science. If Kuhn (1962, 1977) taught us nothing else, he should have taught us that the nature of science changes across time and, by implication, that our archetypes are likely to change as well. Given the equally time-bound nature of our criticism, the archetype approach might be acceptable. That acceptability, however, comes at the high price of begging the question of the standards by which we select those archetypes.

A second limitation of this approach is its implicit legitimation, despite vehement protests to the contrary, of a quasi-objectivist conception of science. I certainly would not want to argue that the rhetoric of science perpetuates the rigid objectivism which, for too long, infected the philosophy of science. It does, however, have the potential for allowing certain objectivist commitments in through the back door. To unreflectively assume a discrete and intrinsic nature of science as the standard by which archetypes are selected is to veer perilously close to the edge of the objectivist abyss. A partial explanation for this condition might be the apparent lack of commitment to a truly constructive conception of rhetorical discourse. In our hesitation to address the "fuzzy" (Rorty 1987) philosophic entailments of a belief in science as a social construction, we retreat to the secure shores of quasi-objectivism. Such a reaction is unduly cautious. We need not ignore the materiality of scientific practice to probe the social construction of science. It is only when we take the social construction position (and its entailments) seriously that we can come to a deeper understanding of the relationship of the material and the rhetorical. The rhetorical dimension is not a unwarranted intrusion into the pristine purity of science; it is an essential element of its very constitution.

A final reason for a skepticism of the archetype approach is perhaps the most important. In essence, it may not be as simple or as accurate as we would like to think. While there might well be a "core set" of relatively unambiguous "scientific" practices, we should not be too hasty in invoking them. Munevar, for example, has argued convincingly that, even with an ostensibly paradigmatic case of science such as space exploration, considerable controversy surrounds its appropriate classification. While most of us (including Munevar) would be inclined to make a presumptive classification of space exploration as science, "the scientific establishment fought the Apollo program on the grounds that it was political showbiz and not science" (Munevar 1985, 420; see also Mitroff 1974). At base, then, there appears to be a compelling need for a more reflective understanding of the nature of science, an understanding that enables our interpretive practices to go beyond both the objectivism of discrete immutable standards and the simpleminded relativism of "everything is rhetorical."

In this chapter, I have argued that the scholarly contributions of the rhetoric of science might be enhanced to the extent that it is willing to move beyond its current penchant for stultifying anxiety attacks and toward an alliance of its traditional concern with practical judgment with a constitutively rhetorical view of the demarcation problem. Through closer and more critical attention to the processes wherein demarcation is enacted through discourse and embodied in variable "ecosystems," we can begin to square our balances of trade with both the intellectual disciplines around us and the larger community that sustains us.

5 THE RHETORICAL CONSTRUCTION OF SCIENCE AND CREATION SCIENCE

The persistence of "scientific creationism" is a curiosity in an age when the epistemic authority of Darwinian, or at least neo-Darwinian, evolution is virtually a commonplace (J. Campbell 1983; Gilkey 1985). Most of us likely want to believe that the antievolutionists' Pyrrhic victory at the 1925 Scopes trial destroyed the public authority of fundamentalist attacks on evolutionary theory (A. Hays 1957; Gilkey 1985, esp. 100–124).

In spite of consistent and unambiguous judicial rejection, however, calls for the inclusion of so-called creation science in American public education continue unabated (Eve and Harrold 1991; Nelkin 1982). While public opinion polls are a risky, incomplete index of the commitments of the phenomenon we call "the public," in this case their findings are at least heuristic. At the height of the 1981 Arkansas "balanced treatment" trial, for example, a national NBC News poll found that 76 percent of Americans favored equal presentation of creationism and evolution in public schools, with 18 percent favoring exclusive presentation of creationism, and a mere 6 percent favoring only evolution (Nelkin 1982).

Although we might wish that these findings were simply anomalous blips on the otherwise clear screen of public commitment to clear thinking, such appears not to be the case.[1] While it would be a bit alarmist to

1. It has been argued that such measures actually index the public commitment to "balanced treatment" or "equal time," rather than support for the particular elements of the creationist agenda (J. Cole 1988; Eve and Harrold 1991, 147ff.). I suspect that such might well be the case. As a practical matter, however, the specific commitment enabling creationism to be included in public school science curricula is less important than the fact that it is included at all. It seems inaccurate to separate creationism as a doctrine from its

suggest that Enlightenment values are on the verge of collapse (at least as a result of creationist influence), signs of the increased influence of creationists have surfaced. On occasion, antievolution has colored deliberations about federal science funding. In 1976, for example, John Conlan (R-Ariz.) sponsored an amendment to the National Defense Education Act to "prohibit federal funding of any curriculum project with evolutionary content or implications"; the amendment passed the House of Representatives on a vote of 222 to 174. While the Senate narrowly rejected the bill, funding for certain National Science Foundation programs was delayed pending a review of their "evolutionary content" (Saladin 1986, 123).

Clearly, it is not likely that such federal initiatives will ever gain widespread support. Of more concern, though, is that creationist pressures are indirectly altering the nature of public science education. Bowing to economic pressure, some textbook publishers have completely dropped references to evolution in favor of "change"; at least one influential textbook reduced its discussion of Darwinism from 2,750 words to 295 words (Woodward and Elliot 1987, 166). Indeed, the most recent California textbook controversy, settled largely in terms favorable to traditional science, nonetheless diluted the wealth of evidence supporting evolutionary theory (Barinaga 1989; Buderi 1989).

As Numbers (1982) has pointed out, creationist efforts have shifted from the legal arena to state and local school boards. It is on that level that creationist efforts appear to be achieving their largest measure of success, as the recent creationism controversy in Vista, California illustrates. Michael Zimmerman recently concluded that 48.4 percent of Ohio's elected officials believed at least mildly that creationism should be impartially taught in public schools. Significantly, he noted that "a large percentage of elected officials showed surprising sympathy for many of the premises of 'creation science'" (1991–92, 39). Affanato's (1986) national study of science teachers found that 45 percent supported the inclusion of creationist doctrine in the classroom, while Zimmerman (1987) reported that more than half of Ohio science teachers and school board presidents did so. The ultimate impact of recent creationist initiatives is perhaps best summarized by Eve and Harrold: "Over a quarter and perhaps as many as half of the nation's high school students get science educations shaped by creationist influence—in

public articulation, which blurs the distinction between substantive content and its political justification. The (perhaps undesirable) conflation of politico-ethical commitments and technical claims illustrates vividly the inherent embeddedness of technical decision making.

spite of the overwhelming opposition of the nation's scientific, educational, intellectual, and media establishments" (1991, 167).

This "overwhelming opposition" seems to make clear that creationism has been professionally discredited as a *science*. This fact has not escaped the attention of critics of the debate. James Hays (1983) concluded that creationism runs afoul of the field-dependent standards of science. As Pine (1984) put it in his neo-Wittgensteinian analysis, creationists are simply playing a "different game." Similarly, Prelli was unequivocal in his assessment that creationists "could not satisfy the minimal tests of reasonable scientific discussion; they were left quite literally without *any* scientific matter for discussion" (1989b, 235).

The work of these and other scholars focus on creationism's clear lack of legitimacy with professional scientific audiences. As many have recognized, however, scientific (and all other) rhetorics undergo significant and consequential transformations in both style and persuasive appeal as they move between various "audiences" (Overington 1977; Prelli 1989b), "spheres" (Goodnight 1982), or "disciplinary frames" (Lyne 1983; Lyne and Howe 1986, 1990). Hence, we are left to speculate on the rhetorical authority of creationism before other sorts of audiences, such as those evincing the high levels of support discussed earlier.

We might be tempted to assign blame for this unfortunate state of affairs to the audiences themselves, in what Ziman (1991) called a "cognitive deficit" account of scientific illiteracy. The woeful educational performance of students in the sciences (as well as the humanities) has long been a topic of concern for professional educators around the country and, in Washington, a series of erstwhile "education presidents." Nonetheless, as Jon Miller has indicated, the "level of scientific literacy in the United States remains low, and . . . the informal science education efforts of recent years have not produced any measurable increase in scientific literacy" (1987, 30).[2]

Root-Bernstein linked scientific illiteracy and the emergence of creationism, commenting that "the layman is ignorant and creationists are preying on that ignorance" (1984a, 16). Jonathan Cole reversed direction but arrived at the same conclusion, arguing that "scientists and educators must clear away a great deal of confusion in the public mind about the true nature of evolutionary science. Confusion that wouldn't be there if it hadn't been created by creationists" (1981, 36). The implicit

2. Of course, the "scientific literacy" literature is large and diverse, including voices dissenting from the assumption that we ought to strive toward universal literacy. For general accounts, see J. Miller 1983; Prewitt 1983; and Trachtman 1981. My goal here is not to enter these debates on their terms, but only as they relate to possible explanations for the nagging "staying power" of creationism.

conclusion of this reasoning is that support for teaching creationism in public schools would wither in the face of more and better science education. While more and better education of whatever variety are laudable goals, it need not follow that creationism would fade away in their wake. To assume otherwise requires making two problematic assumptions: that science has a discrete essence that could be communicated clearly to students, and that students would take, without significant opposition, the clearly communicated essence at face value.

The theoretical discussion to this point, however, gives ample reason at least to defer the first assumption. Reporting on the findings of a Royal Society committee on public understanding of science, Ziman concluded that "the most important finding of the research program is that 'science' is not a well-bounded coherent thing . . . which only starts to be misrepresented and misunderstood outside well-defined boundaries by people who simply do not know any better" (1991, 120).

The second assumption strikes at the heart of the rhetorical concern with audience(s).[3] The widespread assumption that a debate such as that surrounding creationism can (and would) be closed down on the basis of (clearly communicated) technical expertise implies that constituent audiences are largely at the mercy of the technically expert "senders" of the information. In the humanities generally, developments in reader-response theory call this into question (e.g., Fish 1980; Suleiman and Crosman 1980). More directly applicable, however, is the claim that students (or members of a public, more generally) do not react simply to technical content, but to a complex of contextual, institutional, and personal representations of science. As Wynne has argued, "The public uptake (or not) of science is not based upon *intellectual capability* as much as social-institutional factors having to do with social access, trust, and negotiation as opposed to imposed authority" (1991, 116).

If such is the case, we would be well advised to broaden our understandings of topics such as creationism to reflect their inescapable sociality. Such debates, for example, resist closure on purely technical grounds insofar as they implicate broader sets of social practice in which interpretation and evaluation of technical issues are framed by what we would typically consider nontechnical constraints. This broadened understanding recasts nontechnical audiences as active negotiators of technical expertise versus authority. As Wynne puts it, "The public un-

3. I do not assume a myopic conception of "audience" as a discrete component of a speaker-listener dyad. In most (if not all) cases, multiple, active audiences constitute the rhetorical situation under investigation.

derstanding of science represents an *interactive* process between lay people and technical experts rather than a narrowly didactic or one-way transmission of information" (1991, 114).

Here we enter the realm of the rhetorical, of the consequences of discourse. If, as I have suggested, nontechnical audiences actively negotiate the boundaries between technical *expertise* and the *authority* of that expertise to settle questions of a public nature, then the nature of the discourses in which expertise and authority are represented become the primary site of investigation for one interested in accounting for particular controversies such as creationism.[4] Whatever our views on creationism or the larger radical religious right with which it is often aligned, this interpretive stance illustrates graphically the pedagogical and political linkages central to this configuration of the "ecosystem" of science.

Such a perspective has been brought to bear on the controversy previously. Lessl, for example, called attention to the "dialogic" relationship between the creationist "heresy" and the reinvigorated "orthodoxy" of evolution (1988, 31). While dismissing the empirical basis of creationist doctrine, he aptly described the unreflective fundamentalism that characterized the scientific community's response to it. The ways in which the discourses of creationism and evolution constrained each other, however, remain unclear.

Aligning broad cultural analysis with rhetorical criticism, John Campbell argued that "no less interesting . . . is the offense creationism gives to moderns who assumed the scientific world-view so altered the climate of opinion that creationism would naturally go the way of DoDo and the Wooly Mammoth" (1983, 423). Locating an explanation for the appeal of creationism deep within the angst-filled human psyche, Campbell suggests that it reflects the symbolic "management of envy and . . . the eschatological urgency of life against the shadow of the bomb" (435).

4. A related approach might be to situate the creationism controversy as a debate between competing ideologies, for example, between modernism and fundamentalism, or between "equal time" and professional control (Nelkin 1982). As a rhetorical critic, I argue that such ideologies are, in fact, critical topoi in the controversy. Importantly, though, to the extent that such ideologies hold sway, they do so only in and through language. Traditional conceptions of the relationship between rhetoric and ideology construe rhetoric as merely the "handmaiden" of an overarching ideology, merely the means to the end of enacting a preexisting hierarchy of meanings. My perspective, on the other hand, privileges processes of argumentation wherein the contours and relative positioning of competing ideologies are negotiated (though not all negotiators influence outcomes equally). Once so constructed, such ideologies are reconstructed in and through their depiction in rhetorical discourse. On this general process, see McGee 1980; and Lucaites and Condit 1993, xv–xviii.

While the religious foundations of creationism might well offer powerful solace to some sectors of the public, they are but partial explanations for creationism's sustained rhetorical appeal. In my view, attributing too much causality to psychological factors forecloses the potential for meaningful debate on creationism. As a consequence, it immunizes both creationists and traditional scientists from accountability for their rhetorical practices, deflecting the responsibility for interrogating the cultural configurations they enact.

In this chapter, I want to explore a rhetorical account for the public appeal of creation, an account that illustrates the practical consequences of demarcation rhetorics. Whatever the transcendental characteristics of science might turn out to be, the rhetorics of demarcation enacted in this controversy exert powerful (albeit largely unintentional) influence on creationism's public authority. In short, I will argue that creationism endures not only in spite of the response from the traditional scientific community but also, in part, because of that response. My analysis of the creationism controversy between 1975 and 1990 suggests that the demarcational rhetorics of traditional science misconstrued both the scientific pretensions of creationism and the relative insularity of scientific decision making in general.[5] The response proves ultimately unresponsive to the public appeal of creationism, which is grounded in an "empiricist folk epistemology" (Cavanaugh 1985, 1987) and a mistrust of detached technical expertise.

This chapter is organized in three sections. In the first, I offer a reading of creationist discourse which illustrates its quasi-Baconian inductivist demarcation of science.[6] This view of "science," based on an out-

5. This time period was selected primarily because it featured several important court tests of creationist "equal time" legislation (e.g., *McLean vs. Arkansas* and *Aguillard vs. Edwards*) which heightened public visibility of the topic. In addition, the period saw a public controversy regarding the British National Museum's reference to a possible "creationist alternative" in an exhibit honoring the 150th anniversary of Darwin's *Origin* ("Tolerance" 1981, 389). This period also featured a marked rise in the visibility and public advocacy of the Institute for Creation Research, the most prominent creationist "research" facility. Its director, Henry Morris, observed recently that "we have reached more people than ever before" (1991, 3). In its monthly newsletter, *Acts and Facts,* the ICR reports that more than a thousand public lectures or debates on creationism are presented annually by its members.

6. The examples of discourse studied include the major book-length expositions of creationist doctrine published during the period (e.g., Gish 1978; Morris 1980, 1984, 1989; Rusch 1984), the *Proceedings of the International Conference on Creationism,* all volumes of the *Creation Research Society Quarterly* published between 1978 and 1989, as well as all monthly publications of the Institute for Creation Research. The dearth of creationist discourse in traditional "top end" technical journals is a logical consequence of the traditional depiction of creationism as "nonscientific," and hence unsuitable for inclusion in

dated verificationism, underwrites the creationist strategy of attacking evolution and attempts to claim relevance for populist ideographic commitments in the domain of technical decision making. Next, I analyze the demarcation rhetorics advanced by traditional science in response to the rise of creationism.[7] My analysis of the limits and limitations of that response suggests a failure to account for the resonance of the creationist epistemology with the lay (mis)understanding of science and a counterproductive equivocation of technical expertise and social authority.[8] A final section explores the implications of the analysis for more meaningful and humane rhetorics of science which face the difficult tasks of maintaining technical rigor without demeaning competing sets of social practice. As Ziman puts it, "Scientific knowledge is not received impersonally, as the product of disembodied expertise, but comes as part of life, among real people, with real interests, in a real world" (1991, 114). Whatever the ontological standing of our world, the demarcation of what *scientific knowledge* means is less a pronouncement than a rhetorical accomplishment.

Descendants of Dayton: Scientific Creationism

Contemporary creationists, although sharing a particular religious perspective, bear little other similarity to the bumpkins that popular history has long associated with the Scopes trial, by way

scientific journals. Creationists attribute the paucity of creationist articles in such journals to the biases of editors. Such an assertion appears wholly without merit. Scott and Cole (1985) found that only a very few creationist articles have ever been submitted to recognized journals, and the rejection of the few submitted was the product of their failure to adhere to traditional standards of research, rather than creationist content.

7. The discourse studied included all articles, editorials, and letters published in such bridge-level journals as *Nature*, *Physics Today*, and *Bio-Science*. Also included for this study were technical essays published in several volumes designed specifically to respond to creationism, such as *Science and Creationism* (1984), edited by Montagu; *Scientists Confront Creationism* (1983), edited by Godfrey; *Science and Creation* (1986), edited by Hanson; Kitcher's *Abusing Science* (1982); and *But Is It Science?* (1988), edited by Ruse. I also studied volumes published by the National Academy of Sciences, the American Association for the Advancement of Science, and the National Association of Biology Teachers. In addition, all volumes of *Creation/Evolution*, a journal devoted to the creationism debate, were analyzed.

8. I do not call into question what appears to me the obvious superiority of the evolutionary model. The explanatory and heuristic value of Darwinian evolution and its more recent progeny has been demonstrated in a wide variety of disciplines. My concern here is with the failure of the scientific community's response to creationism to come to grips with the public ground for creationism's curious appeal.

of *Inherit the Wind* (Bernabo and Condit 1990). As a theologian (and a plaintiff's witness in the Arkansas trial), Gilkey observes, "The persons who have formulated, elaborated and defended creation science are not preachers ignorant of modern science" (1985, 21). While their contemporary status as "scientists" is, of course, a matter of considerable controversy, most of the leading advocates for creationism have been granted advanced degrees in some scientific discipline from major universities or have worked as "professors or instructors of science" at reputable institutions of higher education (Nelkin 1982, 84). Henry Morris, the director of the Institute for Creation Research, for example, received a doctorate in hydraulic engineering from the University of Minnesota and served for thirteen years as chairman of the Civil Engineering Department at Virginia Tech before becoming president of Christian Heritage College (Morris 1980).

In recent years, the bulk of creationist research has emerged from two related organizations: the Creation Science Research Center and a more publically visible splinter group, the Institute for Creation Research in El Cajon, California. The latter facility, associated with Christian Heritage College, features Morris as director and Duane Gish as associate director. Gish has become perhaps the most visible creationist advocate as a result of his fondness for public debates with traditional scientists (Schadewald 1983). He might be considered the Huxley to Morris' more reticent Darwin in creationist circles (J. Campbell 1989). The major outlets for creationists' technical writings are the *Creation Research Society Quarterly* and books published through Creation-Life Publishers of San Diego. The *Quarterly* is published by the Creation Research Society of Ann Arbor, Michigan, although its editorial board is composed almost exclusively of creationists affiliated with the other main organizations.

The creationists' primary contentions are that scientific evidence supports the Genesis creation narrative and that "academic freedom" demands that such evidence receive "fair and equal treatment" in the public schools. As Gary Crawford explains, "Creationists have attempted to reconcile their beliefs with modern society by using empirical data to prove events described in the Bible" (1983, 4). This "exegesis by science" makes a simple reduction of the current controversy to the myth of the clash between the rigorous thought of scientists and benighted theologians empirically untrue. It similarly suggests that the demarcation practices at issue in this controversy are likely to be especially complex as they attempt simultaneously to exclude and to include various social practices and beliefs.

While there are certain areas of disagreement in the creationist corpus, creationists affirm six general postulates: the sudden creation of the universe, energy, and life from nothing by a Creator; the insufficiency of evolution to explain the development of life and "kinds"; the occurrence of adaptive change only within fixed limits or originally created "kinds"; separate ancestry for human beings and apes; the explanation of geologic history by catastrophism, including a Noachian flood; and a relatively recent origin of the earth, generally described as being between eight and ten thousand years old (Nelkin 1982, 71; Gish 1988; Gish, Bliss, and Bird 1983; Kehoe 1983; Morris 1980).

It is clear, however, that fundamentalist Christian precepts remain at the center of the creationists' scientific program. The preface to the *Proceedings of the First International Conference on Creationism* (1986) closes with the inscription "To God Be the Glory." Similarly, the editorial policy of the *Creation Research Society Quarterly* includes the declaration that "the Bible is the written Word of God, and because it is inspired throughout, all its assertions are . . . true in all the original autographs. . . . The account of origins in Genesis is a factual presentation of simple historical truths." Hence it is the search for and defense of evidence for the so-called simple historical truth of the Genesis narrative that organizes creationist research efforts.

"We Have Met the Enemy . . .": The Creationist Challenge

The contemporary creationism controversy is often viewed as the most obvious example of pseudoscience attempting to find a public hearing and following. In short, it is generally assumed that the explicit declarations of creationists to be "doing science" are simply self-conscious attempts to distract attention from their real purpose, which is not at all scientific but rather religious. (See, e.g., Gallant 1984; Kitcher 1982; K. Miller 1984; Moore 1983; Murray and Buffaloe 1983.)

What is contested in this conventional wisdom is not the religious orientation of "scientific creationism." Indeed, the creationists are frank about that (Gish 1988; Morris 1980). What is suggested is that creationists really have no aspirations of "doing science" but want only to coopt the terms and epistemic authority of science so as to construct a fundamentalist hegemony in public education. To a very large extent, the traditional evolutionary (and public) response to creationism has been simply to scoff at what are taken to be the obvious lack of sincerity and transparent deceits of creationist charlatans.

While creationists' frequent misquoting of traditional scientists makes this a tempting characterization, a dismissal of the creationist program as no more than a deliberate exercise in obscurantism is at best inaccurate and ultimately counterproductive. Careful analysis of creationist discourse reveals the construction of an image of science that organizes the creationist attack on evolution and, more significantly, parallels the dominant public understanding of science and the processes of scientific research. This is not to say that the image of science constructed therein is one of equal legitimacy or empirical validity to the reigning paradigm. Indeed, it is one that has been generally discredited, and for very good reasons. Nonetheless, that image underwrites the public demarcation rhetorics of creationism.

Apparently unencumbered by the quantum revolution and the past century of the philosophy of science, creationist rhetoric reflects, as Marsden (1984) and Kehoe (1983) have suggested, the epistemic commitments of the first "scientific revolution," grounded in seventeenth- and eighteenth-century inductivism. While programmatic epistemic statements are rare in creationist literature, it is clear that there is a close alliance with caricatures of Baconianism, tempered with the Scottish Common Sense Realism of Thomas Reid (Bozeman 1977; Kehoe 1983; Marsden 1984).

True science, from this perspective, abandons so-called metaphysical flights of fancy in favor of close empirical observation and strict processes of induction from those observations. As Bacon wrote in his *Novum organum,* we should seek "not pretty and probable conjectures, but certain demonstrable knowledge." The human mind, Bacon insisted, "should not be left to take its own course, but should be guided at every step . . . as if by machinery" (1900, 4.40). As Marsden puts it, "While some judiciously conceived hypotheses might be entertained to guide generalization and experimentation, [those] . . . incapable of verification by observation are beyond the realm of true science" (1984, 97).

In addition to this superficial epistemological heritage, creationists appear similarly indebted to Reid's commonsense philosophy which holds that human beings are compelled by nature and common sense to accept certain beliefs, including the existence of the external world and the consistency of an individual's perceptions of it (Grave 1960; Kehoe 1983). In his *Enquiry into the Human Mind* (1970), Reid observed that "sensation carries with it an immediate belief in the reality of the object, and this immediate certainty supplies us with a criterion of truth" (121). In "Essays on the Intellectual Powers of Man," he maintained that "the Supreme Being intended that we should have such knowledge of the material objects that surround us . . . and he has admirably fitted our powers of perception to this purpose" (1966, 146).

From this perspective, science is portrayed as an egalitarian activity, indeed a universal human capacity, and this is a view quite at odds with prevailing cultural assumptions regarding science as "uncommon sense." Human beings, by their very natures, are said to be necessarily committed to certain beliefs, such as the existence of the external world. Since such beliefs are said to be universal, arguments from evidence of careful observations can attain as much objective (read: scientific) certainty as the human race can ever hope to attain (Grave 1960). In Reid's terms, "the information of the senses is as perfect, and gives as full conviction to the most ignorant as to the most learned" (1966, 147).

Significantly, this outmoded philosophy of science is not restricted to the occasional creationist rhetor. Cavanaugh (1985) argues that alarming levels of scientific illiteracy among the general public have reinforced this "empiricist folk epistemology." While I would not ascribe responsibility to a totalizing "illiteracy," preferring also to consider mediated representations of science and so on (LaFollette 1990; Nelkin 1987; Taylor and Condit 1988), it seems clear that, while this naive verificationism has long been rejected in most intellectual circles, it still carries significant public authority (J. Miller 1987; Prewitt 1983).[9]

That creationists adhere to such an epistemology *and* definition of science is quite clear. Morris, for example, argued that "science is *knowledge* and the essence of the scientific method is experimentation and observation. Since it is impossible to make observations or experiments on the origin of the universe . . . the very *definition* of science ought to preclude use of the term when talking about evolution" (1974, 249). Writing in the *Creation Research Society Quarterly*, McGhee argued that "the bedrock of real science is trust in the purity and accuracy of trained autonomous observation and reason" (1987, 138). Morris and Parker, emphasizing a naive empiricism, insist that "nothing is really easier for scientists and just ordinary people than finding and recognizing evidence for creation. . . . *Just use the ordinary tools of science: observation and logic*" (1982, 2).

This rhetorical allegiance to outdated empiricism is not limited to creationists' broad epistemological pronouncements. It is also manifested clearly in what passes for creationist research, such as that presented at the 1984 International Conference on Creationism. For exam-

9. There is also the question of whether creationists even accurately represent Bacon and/or Reid. Revisionist readings of Bacon, for example, minimize his commitment to "certainty" and soften the traditional view of his rigorous induction. See, for example, Urbach 1987. While the true "essence" of Baconianism or commonsense philosophy is a topic better left to others, my reading suggests that creationists enact a rather caricatured version of both.

ple, in their analysis of deformation features in Cambrian sedimentary rocks, Austin and Morris conclude that "creationist predictions concerning the relationship between sedimentation, tectonics, and cementation more closely correspond to *reality* than do evolutionists' predictions" (1986, 15).[10] Lucas, in his efforts to craft a theistic "unified theory of modern science," challenges the theory of general relativity as "logically false and inapplicable to the data and phenomenon of the real world" (1986, 127, 132). On the other hand, he claims that his creationist electrodynamic model for charged elementary particles is "in better agreement with reality" (128). As creationists would have it, "The facts speak for themselves, and . . . open-minded, open-hearted readers will be convinced" (Morris 1989, 16).

While most of us have assigned such an epistemology to the dustbin, creationists appear to revel in their "outdatedness." In fact, Rusch (1984) invokes the "law of priority," arguing that, since Baconianism predates contemporary philosophies of science, it reigns as the proper model for scientific investigation. As he puts it, "Under the law of priority, there would be the requirement that a new name be found for what passes as science today, and according to priority, only scientific creationism would be *Science*" (16).

While the argument from priority seems specious, that alone does not make creationism nonscientific or pseudoscientific. While the creationists' notion of science as grounded in the bedrocks of reliable observation and strict induction resonates with the literal and authoritarian nature of their theology, this need not entail that the former is simply designed to mask the latter. It is not necessary to impugn the sincerity of creationists in order to challenge their empirical and theoretical claims, not to mention the reactionary political agenda that often accompanies them. Indeed, an analysis of creationists' demarcation practices is crucial to understanding what many people take to be the utterly nonsensical nature of their public and technical discourses. An analysis of the implications of this Baconian demarcation should serve to demonstrate this point.

It is clear that the creationist notion of science is at odds with contemporary scientific practice, not to mention most of the philosophy and sociology of science written in the last twenty years. It is not im-

10. In brief, this controversy concerns the implications for certain soft sediment deformation features in thick sequences of strata for traditional assumptions regarding geologic age. Austin and Morris (1986) argue that such sequences could not have escaped lithification if they had been buried for millions of years, as traditional geology suggests. Evidence of deformations they consider characteristic of much younger sequences are cited as evidence of a recent, catastrophic account of geologic stratigraphy.

portant here, however, that creationists may not (indeed, cannot) truly operate as utterly naive inductive empiricists. The concern here is with the naive inductive empiricism to which they pledge intellectual allegiance. It stands in stark contrast to the implicit working truths which organize scientific practice and our understandings of the epistemic products of that practice. Christopher Nelson, emphasizing this contrast, notes that for evolutionists "scientific knowledge is fundamentally uncertain. . . . No amount of agreement, not even perfect agreement, between predictions and data ever justifies our believing a hypothesis is certainly true" (1986, 129; see Popper 1959). Botanist Roger Alexander maintains that "the creationists' argument suggests that a fact is something that, once discovered, is kept forever. . . . Not so. Nothing is irreversibly factual. Any fact may turn out to be not a fact at all" (in Kottler 1983, 33). Indeed, it might be argued that, in modern evolution-influenced sciences, evolutionary postulates serve as fundamental organizing frameworks for research rather than discrete observations. As Toulmin noted, "Scientific discoveries are typically arrived at not by generalizing from preexisting *facts* but by providing answers to preexisting *questions*" (1982, 95).

What I take to be important about this contrast is that, at least for sincere creationists, the current controversy is indeed a *scientific* one. Among other things, then, this controversy involves competing demarcations of science. Hedtke, adopting a curious notion of natural philosophy, argued that "since evolution was established according to natural philosophy, rather than the more rigorous natural or exact science guidelines, the controversy centers *first and foremost* on natural philosophy vs. natural science. . . . The antagonists are not, as commonly perceived, science and theology. . . . The issue [lies] primarily within science itself" (1979, 89). Even more pointedly, Morris rejects the a priori distinction between the sacred and the scientific, insisting that "the Biblical world outlook *is* the scientific outlook. . . . [The universe's] processes and systems are reliable and intelligible, operating in accordance with fixed laws that can be discovered and used" (1989, 304). Viewing the current controversy from the perspective of the majority of creationists orients attention away from relative sincerity of motives (or lack thereof) and toward the competing (and relative merits of) demarcations of science.

This orientation entails understanding the explicit creationist repudiation of subjective personal intervention in the scientific process. Creationists claim that evolutionary theory requires the active intervention of its adherents in the form of "fitting the facts" to a taken-for-granted theory. In his creationist discussion of the philosophical underpinnings of geologic evidence, Strickling argued that "subjectivity is seen in [evolutionary] uniformitarianism in its frequently documented rejection of even contemporary hard scientific evidence that contradicts its tenets" (1979, 98). Similarly, Hedtke maintained that evolutionists (including Darwin) tend to "develop an *a priori* hypothesis and then, by a variety of machinations, fit all facts into the hypothesis or monger in subsidiary hypotheses to explain away conflicting facts" (1979, 89).

Technical illustrations of this alleged tendency abound in the creationist literature. Overn and Arndts, advancing the common creationist attack on radiometric dating techniques, maintain that "a very clear mathematical trick called 'isochrons' has more recently been used by geochronologists to delude themselves into thinking that they are able to produce rigorous proof for old-age rocks in radioactivity data" (1986, 167). Morris and Parker, challenging traditional explanations of "missing" stratigraphic levels, insist that "evolutionists have coined a term to deal with the problem: *paraconformity*. . . . Creationists don't need the term paraconformity. Creationists can simply accept the physical evidence as it's found" (1982, 133). While welcoming the alleged challenge to the "big bang theory," DeYoung (1992) criticized recent developments in "plasma cosmology" (Lerner 1991) as requiring arithmetical gymnastics. Referring to evidence produced in laboratory plasma chambers, DeYoung argues that "the unwarranted extrapolation of plasma properties throughout deep space . . . is similar to modeling galaxy formation on a computer, and then declaring that the computer simulation is reality. Must creationists be the only ones to blow the whistle on this poor logic?" (1992, iv).

Of course, it can (and probably should) be argued that such criticisms are, at best, ingenuous and, at worst, an offensive form of intellectual hypocrisy. To suggest that creationists do not manipulate data, with the effect of advancing their own religious program, is simply mistaken (J. Cole 1981). Recall, however, that the point at issue here is not that creationists engage in precisely the same behaviors which they label "nonscientific" for evolutionists. Consistency of application of a given implicit demarcation criterion is a topic for others to consider. My concern is with how this implicit demarcation criterion structures the evaluative discourses of the creationist community.

At base, the implicit portrayal of science constructed in creationist discourse serves to warrant (for them, at least) the ostensibly damning criticism of evolution as unjustifiably speculative, hence nonscientific. Gish maintained that "for a theory to qualify as a scientific theory, it must be supported by events, processes, or properties which can be observed. . . . The general theory of evolution fails to meet these criteria"

(1978, 13). In a tactic quite common among creationist advocates, Morris reduced the complex issues at stake to simplistic generalities. He argued that "evolutionists seem to be saying: 'Of course, we cannot really *prove* evolution, since it requires ages of time [but] you should accept it as a proved fact of science.' Creationists regard this as an odd type of logic, unacceptable in any other field of science" (1980, 16). Austin and Morris, responding to evolutionary accounts of sedimentary strata, proclaimed that "scientists should be more interested in explaining what *has* been found than in defending . . . what *has not* been found" (1986, 3). Chaikowsky perhaps best clarified the incredulity with which strict Baconian creationists regard the speculative scientific aspirations of evolutionists when he offered the following bit of whimsy in describing the intellectual quandary facing the evolutionist:

> As I was sitting in my chair,
> I knew it had no bottom there,
> Nor legs, or back, but I just sat,
> Ignoring little things like that.
> (1979, 174)

Acording to the creationist story, the evolutionist must necessarily find herself or himself in just such a precarious position: without any visible support and dependent on nonobservational beliefs to prevent crashing to the cold, hard floor of "scientific fact."

This initial act of demarcation has important implications for the consequent critique of evolution. Initially, and perhaps most fundamentally, such a demarcation "allows" the creationist to maintain that evolution, despite its overwhelming evidentiary support, is not and cannot be *more* scientific than creationism. Seeking to reclaim scientific legitimacy for creationism's admittedly teleological account of human origins and development, Woodmorappe argued that the unempirical nature of evolution, at the very least, made the competing theories scientific equals. He maintained that "it is high time that scientists recognize that a teleological explanation can be as scientific as a (however unempirical) materialistic scenario for the unobservable past" (1979, 209). Morris and Parker concluded that "both the Creation Model and the Evolution Model are, at least potentially, *true* explanations of the scientific data related to origins, and so should be continually compared and evaluated in scientific studies" (1982, xiii).

This superficial leveling of evolution and creationism carries significant implications. Most notably, it suggests that, since neither program is capable of sense verification, the two deserve equal treatment in research and pedagogy. This meshes neatly with the commitments to "equal time" or "balanced treatment" that are sustained by public representations of the creationist agenda (Taylor and Condit 1988). The argument that creationists simply invented what looked like a scientific justification to mask their real intent in the wake of the Scopes debacle is unconvincing to me (Alexander 1978). That this philosophical justification has been almost completely repudiated by the academy does not disprove that it is a rhetorical product of the creationist view of science, nor does it deny that the justification carries, perhaps unfortunately, a measure of public authority.

An additional implication of this demarcation of science was its implicit warrant for the insertion of the Creator in the "scientific" creation story. The creationist perspective on science, grounded in the infallible operation of the senses and the fallibility of contemporary accounts of origins, is constrained to posit what it takes as the least implausible extrapolation from sense data. To the extent that the universe is said to show "overwhelming evidence" of design (Morris 1980; Woodmorappe 1979) and that the evolutionary scheme is limited to positing an unknown origin and unobservable methods of progression, it is consistent for the creationist to posit benevolent acts of creation as a more likely explanation of origins and the current state of nature. (Intellectually vacuous perhaps, but theoretically consistent, nonetheless.) Chaikowsky recognized that a Creator was akin to heresy for adherents of the evolutionary perspective. An "honest evolutionist," however, must be compelled to admit that he or she cannot offer a definitive (read: observable) explanation of the "introduction of matter" (Chaikowsky 1979, 174).

The creationist, on the other hand, can, even must, offer precisely such an explanation: the ostensibly undeniable commonsense observation of the natural order, it is argued, lends more direct credence to an active creation than to an unknown act of creation and what creationists portray as "haphazard development by willy-nilly chance" (Morris 1974, 178). Baumgardner concedes that "it seems evident that the Flood catastrophe cannot be understood or modeled in terms of time-invariant laws of nature. Intervention by God in the natural order during and after the catastrophe appears to be a logical necessity" (1984, 24). Reflecting on his reading of the fossil record, Gish pondered, "*What greater evidence for creation could we give than this sudden outburst of highly complex forms of life?* . . . The rocks cry out, 'Creation!'" (1985, 69). John Morris gloated that "creationists have another advantage. Even though we can't 'study the past' we can study the record of One who was an active eyewitness throughout the past" (1992, d).

The relative historical veracity of such an extrapolation is not being defended here.[11] I merely suggest that the view of science rhetorically demarcated in creationist discourse constructs a symbolic reality in which creationist knowledge claims are given a measure of internal consistency. It is this internally emergent consistency that allows creationists to warrant their pairing of the supernatural and the ostensibly scientific into their unified and unifying worldview.

The creationists' modern use of this demarcation rhetoric might also be read as indirectly legitimating their particular ordered view of the world and its attendant social implications. Consider the creationist view of secure scientific knowledge as necessarily grounded in commonsense observation of a stable empirical reality. Given that such commitments are said to be natural constituents of human nature, it follows that scientific knowledge would be equally accessible and secure regardless of historical context. As Marsden argued, "A feature of such views important for understanding them is that they emphasize stability and uniformity of human knowledge across time and culture" (1984, 100). To borrow Rorty's (1979) phrase, true Baconians and their creationist counterparts construe our scientific perceptions as "mirrors of nature."

On a conceptual level, the stability thought to be inherent in scientific knowledge is directly contrasted with the intrinsic "open-endedness" of traditional evolutionary science. Indeed, the very notion of evolution suggests a progressive adaptation of species and, as a consequence, of the methods for their apprehension. This evolutionary insistence on the ultimate uncertainty of science is at odds not only with the creationist Baconian construction of science but also with the lay understanding of scientific certainty. Nelkin argued that "for the creationists, order is a fundamental value; they are distressed by the flux, uncertainty, and doubt inherent in science and basic to the theory of evolution" (1982, 42). By contrast, the creationist view of science is thought to offer a consistent and coherent system that fully explains the natural world. Invoking memories of William Jennings Bryan's "rock of ages vs. ages of

11. It is at this point that the critic faces a difficult decision—to repudiate creationism and its attendant politics, too often by invoking precisely the sort of totalizing claims of "it's not scientific" that become problematic, or to resist all such definitional gambits with the unwelcome possibility of being construed as granting creationism equal intellectual and political value. I want to reject such a dichotomous analysis, however, in favor of a Latourian stance toward the stability of knowledge claims. It seems to me that many of the particular claims unified within and by evolution are sustained by myriad practices and are recognized as central to the practice of science. As a consequence (and ontology aside), it makes little sense to pry open every potential "black box"; we simply must be more judicious in choosing those left intact and in justifying those choices.

rocks" proclamation at the Scopes trial, a biologist at the Institute for Creation Research observed that "science can't be trusted but God can. . . . [Evolutionary] data is so permanently incomplete that it is hardly a good place to sink an anchor for anything to do with eternity" (in Nelkin 1982, 42).

Accentuating the rhetorical appeal of this ostensible stability is the creationist implication that science, as the disciplined exercise of commonsense observation, is an egalitarian social practice, available to all those who would participate properly. M. Bowden's creationist reading of the human fossil record presumes that "the layman's judgment can be as valid as that of the expert," given "the presentation of *all* the relevant evidence" (1981, 1). Austin and Morris, recalling their field research at Split Mountain, California, remarked that "every observer, regardless of education, is struck with awe and wonder upon viewing the Split Mountain Fault and the unusual deformation features associated with it" (1986, 6).

Creationists exploit this surface-level egalitarianism with their oft-repeated charges that their "outsider" status vis-à-vis the traditional scientific community is a function of the myopic oppression of an elitist cadre of humanist evolutionists. Portrayed as denizens of the same ivory tower thought to harbor such miscreants as absent-minded professors engaged in abstruse philosophizing and/or tenured radicals with their politically correct proselytizing, traditional scientists appear to be keepers of their own exclusive dogma. Bemoaning what he saw as unfair treatment of creationists at the sixty-third annual meeting of the Pacific Division of the American Association for the Advancement of Science, Gish begrudgingly agreed to "take one of the two seats on the back of the bus reserved for creationists at this meeting" (1984, 26). Commenting on the denial of accreditation by California's superintendent of public instruction for graduate programs at the Institute for Creation Research, *Acts and Facts* indicated that "Honig and company had other designs. . . . [They were] determined to send an elite team of evolutionary scientists to destroy the school once and for all" ("Victory" 1992, 2).[12] Creationist lawyer Richard Turner ridiculed evolutionists because "they get up on the stand, and act as if their very lives were being attacked. . . . They're pompous and arrogant, just the kind of people the First Amendment was written to protect us against" (in Broad 1981, 1332).

12. After protracted controversy, the California Department of Education dropped its efforts to deny accreditation for creation science programs at ICR. ICR now boasts full accreditation from the Transnational Association of Christian Schools ("Creationist Schools" 1992, i).

Beyond its broad public implications, this understanding of science as a stable stock of knowledge furnishes the creationist a closer alliance with his or her religious beliefs and professional activities. The epistemic monism which characterizes the creationists' science is "mirrored" in their monistic fundamentalism. The alleged literal truth of the Bible provides a stable foundation of beliefs which, it is thought, the scientific enterprise is properly organized to explicate. Indeed, Morris insists that the doctrine of creation is the basis of Christian faith itself, remarking that "the doctrine of origins . . . is the foundation of every other doctrine. That, of course, is why God placed His revelation of origins in the first chapter of the Bible. Everything else in the Bible and in history is built on this foundation" (1980, 26).

While there are some uncomfortable implications of this position, I suggest that the resonance of creationists' epistemic commitments and Christian fundamentalism belies the easy demarcation of science *from* religion. The professional activities of the creationist are oriented toward explicating their inescapable interaction. For purposes of this argument, I need merely conclude that it is unwise (and, I think, inaccurate) simply to write off the creationist as one who uses the label of science to hide religion. For a creationist, that a priori demarcation is untenable. That is not to say that the creationist is correct. It is merely to say that the creationist rhetoric of demarcation creates a symbolic universe in which two profound social practices, the religious and the scientific, can be reconciled. Whether or not the particular way that creationists reconcile them is a useful or constitutional one will be left for another time.

Perhaps the most important rhetorical implication of the creationist demarcation of science is its implicit sanctioning of the most common method of attacking the evolutionary paradigm: the use (and abuse) of apparently disconfirming examples to discredit evolution. Creationism is far less than a constructive program of independent scientific conclusions that cumulatively yield conceptual and empirical support for the creation narrative; rather, it functions primarily by pointing out apparent anomalies which current evolutionary thinking has yet fully to explain. Indeed, the very notion of "dual models" of origins and development made popular by creationists (e.g., Morris 1980) implies that creationism and evolution exist in a zero-sum relationship. In essence, evidence against one view is taken to count as evidence for the other. While such a strategy has been soundly repudiated as misleading, it is made sensible for creationists through its "fit" with their rhetorics of demarcation. Recall that "true science" is said to be based on direct empirical observation of an stable, unchanging reality. By consequence, the "facts" don't change; if an allegedly scientific account is unable to match its predictions with those "facts," then that account is said (by creationists) to be inaccurate, and hence unscientific. Accordingly, for a creationist, the strategy of pointing out counterexamples is more than merely nitpicking.

This orientation toward matters of explanation and proof is enacted in the most common creationist attack on the reigning evolutionary paradigm: the so-called gaps in the fossil record. While I have alluded to it previously, I want here to describe more directly the creationist pattern of argument on this issue. My goal is not to provide a full-blown analysis of the controversies themselves but rather to illustrate how the creationist demarcation of science is played out in specific evidential contexts.

The title of Gish's book *Evolution: The Fossils Say No!* (1978) concisely captures a crucial pillar of the creationist attack on evolution: the alleged inability or failure of paleontological evidence to confirm the claims of evolutionists. In various forms, creationists argue that the fossil record is incomplete, containing allegedly inexplicable gaps where steady progression of fossilized evidence should appear, forcing traditional scientists to invent unjustifiably speculative explanatory "bridges" over nature's empirical "walls" (J. Campbell 1986). Gish argued, "If life arose from an inanimate world through a mechanistic, naturalistic evolutionary process via increasingly complex forms into millions of species that have existed and now exist, then the fossils actually found in the rocks should correspond to those predicted on the basis of such a process" (1978, 33).

What is of special interest here is the creationist understanding of the term *correspond*. Following from the creationist demarcation of science, it suggests the necessity of providing clear observable evidence of the past existence of now extinct species, that is, of providing a complete fossil record. Gish asserted that "it seems clear, then, that after 150 years of intense searching a large number of obvious transitional forms would have been discovered *if* the predictions of evolution theory are valid" (1978, 49). Kaufmann concurred, insisting that "if evolution is really supported by the fossil record, intermediate fossil forms should be found in massive numbers" (1987, 81).

Not surprisingly, a primary creationist strategy is to point out cases in which evidence of transitional forms is incomplete. For example, Gish, in his discussion of the evolution of crossopterygian ichthyostegids, notes that a considerable fossil gap exists. He maintains that such a gap "would have spanned many millions of years" and should reveal "a slow gradual change of the pectoral and pelvic fins of the . . . fish into the feet and legs of the amphibian . . . and the accomplishment of other transformations required for adaptation to a terrestrial habit" (1978, 78–79). Similarly, Kaufmann argues against what he takes to be the inanity of the Darwinian claim of the evolution of reptilian life-forms *(Protoavis)* to avian forms *(Archeopteryx)*. Citing the unstable correspondence between their fossilized remnants, he suggests that "one conclusion is certain. Neither of these classic examples of evolution is really compatible with the claim that living birds are the end product of a process which started with cold-blooded reptiles" (1987, 185). The fact that the evidence on these (and other) transitional forms is unclear leads Gish to gloat that such evidence "rather strongly contradicts the evolution model" (1978, 68). At the 1984 meeting of the American Association for the Advancement of Science, Gish maintained that "at the very best, the evolutionist has three or four or five suggested transitional forms which he comes back to again and again" (1984, 33).

Given the creationist demarcation, this alleged paucity of evidence is tantamount to the "descientizing" of evolution. As the creationist would tell the story, evolutionists substitute antiempirical speculation for properly scientific observation and strict inductive extrapolation.

This critique of the evolutionist reading of the fossil record as speculative serves to ground creationist claims to explain more adequately the available (and observable) evidence that the fossil record does provide. Recall the creationist insistence that evolutionary changes are restricted to changes "within kinds," that is, within originally created life-forms. This belief entails far fewer transitional forms and appears elastic enough simply to label any extant fossil form as a "kind." Rusch indicated that a "creationist . . . holds that there are basic kinds that seem to remain intact, varying within the basic kind but not across *kind* lines" (1984, 36). Gish, Boardman, Koontz, and Morris pointed out the significance of the fossil record to the "theory of kinds." They contended that "when all is said and done . . . examples of present day evolution that are commonly cited in textbooks are nothing but relatively minor variations within originally created kinds" (in Gould 1983, 151).

It should be noted that creationist authors are masters at decontextualizing, manipulating and distorting the reasonable admissions of evolutionists not to have discovered a complete fossil record. Indeed, such a complete record is in principle unattainable. Raup and Stanley insist that "any study of fossils or use of paleontological data must be based on a clear understanding of the strengths and weaknesses of the record" (1978, 8). Among the weaknesses of the record is that "fossilization is far from being the inevitable destiny of a past organism" (Kitcher 1982, 107). Dead organisms can be destroyed before fossilization by any of a number of natural means. To claim that absence of a particular transitional form demonstrates that such a form never existed seems, at best, specious.

Therefore, creationist discourse does articulate a historically grounded view of science. In its insistence on empirical observation, strict induction, and commonsense realism, creationism demarcates science in a way that serves actively to constrain evaluations of the evolutionary paradigm. I have also argued that a singleminded and a priori dismissal of those creationist claims as simply a subterfuge to gain respectability for an exclusively fundamentalist Christian doctrine is a form of critical myopia, and one that leads to ironic and unfortunate consequences. To understand the symbolic world in which creationists conduct their research and in which they defend it before public legislative and judicial forums requires that we recognize the epistemic grounding of the creationist discourse itself.

While the claims of creationism might well seem arcane, misguided, or just plain silly to many of us, it is unduly elitist to leave it at that. We cannot afford, as Burke noted, to offer a few adverse attitudinizings and call it a day. Those committed to a neo-Darwinian view of natural history do not have a monopoly on clear thinking and/or intelligence. The epistemic and rhetorical commitments of the competing camps are so radically different that both groups seem reduced to repeating threadbare and often misleading shibboleths so as to demean the other, not the least of which is the claim that creation (or evolution) is not *science.* As Marsden concluded, "Each party thinks that the members of the other are virtually crazy or irremediably perverse. Neither thinks that the other is doing science at all" (1984, 109).

All that aside, one is still left wondering why the clear evidential superiority of evolution hasn't won out—yet. Perhaps it is possible to attribute expertise to a rhetor while simultaneously denying her or his authority to foreclose public debate on the broader implications of the technical claims in question. In short, the persistence of creationism is inexplicable only to the extent that we characterize the creationism/evolution debate as exclusively a technical one (e.g., J. Hays 1983). Willard (1990), for example, has conceded that "if the *creationism vs. evolution* debate carries political, legal, constitutional, pedagogical, and moral implications, then the advocates of either side possess no special claims on the deliberative processes that the public sphere should employ in evaluating their dispute" (150). Using the ecosystem metaphor, as the complexity of its interconnections increases, the presumptive primacy of any given constituent decreases, though its potential primacy remains. In the present case, then, the key is to develop rhetorics that redeem that potential through argumentative engagement.

The answer to our conundrum might be precisely that the rhetorical demarcation of science is mishandled when it is reduced to the recitation of ahistoric standards. The case of creationism suggests that, even in what are generally taken to be unequivocal cases of pseudoscience, the grounds for so classifying particular belief systems are more complex than generally thought. A rhetorical perspective on demarcation offers a reflective understanding of that complexity.

The following section of this chapter will begin the analysis of the interest-driven demarcation practices of evolutionists as they redefine science in response to the creationist threat.

The Deployment of Demarcation: Evolutionists on Creationism

Empowered by its growing political currency (Nelkin 1982), the creationist discourse embodied a series of compelling challenges to the epistemic and cultural authority of contemporary science in general and evolutionary theory in particular. Sensing some rhetorical success, one creationist declared that "we are making considerable impact. . . . the alarm is being felt by evolutionists . . . they are feeling a need to go out and meet our challenge" (in Nelkin 1982, 43). The unquestioned dominance of a particular understanding of science and one of its central theoretical offspring was at stake in the creationism controversy.

The first major interest threatened by the cultural emergence of the creationist discourse involved the indictment of a fundamental precept of modern (especially natural) science. Darwinian evolution, as a research topic, metatheoretical and methodological premise, serves as an organizing constituent of contemporary science. Gould noted that "evolution is one of the 'great ideas' of science. . . . It is a tenet without which science as we know it could not continue to exist" (1981, 35). The implications of evolutionary thinking on a wide variety of scientific disciplines are very clear. Dobzhansky maintains that "nothing in biology makes sense except in the light of evolution. . . . Without that light, it becomes a pile of sundry facts—some of them interesting or curious but making no meaningful picture as a whole" (1983, 27). Kitcher argues that evolutionary precepts extend to the totality of the scientific enterprise. He contends that "evolution is intertwined with other sciences, ranging from nuclear physics and astronomy to molecular biology to geology. . . . There is no basis for separating the procedures and practices of evolutionary biology from those that are fundamental to all sciences"

(1982, 4–5). The increasing public and legislative currency of creationism represented a threat to this epistemic order. William Mayer, director of the Biological Sciences Curriculum Study, lending some credence to Lessl's contention that evolution has risen to a level of "root metaphor or world view," argued that "the whole nature of science is under attack. . . . It's not just biology, it's all of science" (in Lewin 1981, 635).

Lessl observed, "Just as in countless religions where human activities have divine archetypes in the eternal realm, we find that the activities and experiences of the scientist have an eternal paradigm . . . evolution" (1985, 177). While there is considerable disagreement within traditional scientific circles regarding the particular mechanisms by which evolution is accomplished (e.g., Gould 1983), it is difficult to overstate the intellectual authority of its general principles. Nobel laureate P. W. Medawar concluded that "the evolutionary hypothesis makes sense of the natural order. . . . In biosystematics and comparative zoology, the alternative to thinking in evolutionary terms is not to think at all" (in Moore 1983, 13). In this sense, then, the rise of creationism can be read as indicting a central conceptual and methodological commitment of contemporary science. That indictment ultimately necessitated a rhetorical response which served to (re)construct what it meant to do contemporary science.

A second communal interest at stake in the creationism controversy involved the organs of authority traditionally assumed to control and/or sanction scientific practices, practitioners, and their proclamations. This is a matter of no little significance. As Whitley argued, "Freedom to act and speak in the name of 'science' is a privilege and not a right" (1985, 17). Implicit in this internal scheme of things is the notion that the allocation of that privilege is dependent on an authorizing decision of the scientific community itself. To the extent that only *scientists* do *science*, then, the rhetorical classification of given social actors as scientists makes the demarcation question a central (and deeply contextualized) exigence.

In this sense, creationism and the creationists represented a singularly awkward anomaly for the authorizing community. As I noted earlier, the contemporary creationism movement is not grounded in any explicit dichotomy between science and religion. Indeed, creationist advocates have been granted the most public imprimatur of science: advanced degrees from generally respected universities (Morris 1980; Nelkin 1982). Creationists, of course, are not reticent about proclaiming their "legitimate" status in a tactic that Kitcher aptly labels "credential mongering" (1982, 178). Those credentials presumably sanctioned the scientific legitimacy of the degree holders because, as Dolby pointed out, "formal scientific institutions limited themselves to the higher parts of cultural hierarchies by setting *minimum* standards for participants in their activities" (1982, 273). However indirectly, those meeting the standards were included in the symbolic world of science and scientists. That creationists could potentially take up residence threatened the very foundations of that world.

In addition to the theoretical paradox represented by creation scientists assailing the foundations of science, creationism represented a very practical threat to the system of control through which rhetorical legitimacy is allocated. Nelkin argued that "control is maintained through informal internal communications and through a peer review system that determines research funding and the acceptance of papers by journals" (1977a, 273). Creationists, however, claimed publicly to be "scientists," and they adopted at least some of the language and forms of science, raising the possibility that the system of control was unable to constrain the dissemination of what was taken as a "nonscientific" doctrine, or that the system of control was utterly irrelevant to everyday practice and hence could not claim to vouch for science at all. In either case, an important internal interest of the dominant scientific community was challenged.

A final threat to internal cognitive interests was perhaps the most fundamental, insofar as it questioned the prevailing conception of rationality around which scientific practice is organized. In short, the rhetorical emergence of the creationist discourse questioned the foundational epistemological assumptions of evolutionary science. Evolution is widely considered the cornerstone of contemporary science. As such, its epistemic status within the scientific community is hallowed. Far from being simply "another theory" (Morris 1980), evolution has approached a level of acceptance which belies the properly contingent (albeit vastly well-documented) nature of its constituent knowledge claims. Wiley argued that "there is *no* controversy over the phenomena of evolution" (1981, 730). Eldredge was equally succinct: "We scientists should stop mincing words: evolution is a fact" (1982, 121). Even those scientists (and there were many) who expressed misgivings over such bald paeans to the epistemic quality of evolution did so less on epistemic grounds and more on grounds of political expediency (Kottler 1983, 31-35).

While it is likely that many practicing scientists would reject, on philosophical grounds (if they consider them at all), the notion of scientific truth as an absolute, such assumptions play a central role in the conduct of scientific practice. Broad and Wade observed that "the myth of science as a factual/logical process, constantly reaffirmed in every article, textbook, and lecture, has an overwhelming influence on scientists' perceptions of what they do" (1982, 126). Lessl argued that "while stating that evolution is a *fact* does, in one sense, violate the presumptions of the community . . . a factual treatment of evolutionary theory is rhetorically important to the cosmic [scientific] myth" (1985, 178). In this sense, the "truth" of evolution is a crucial presupposition of science, without which, as Gould noted, "science as we know it could not continue to exist" (1981, 35).

Such metaphysical debates might best be left to professional philosophers. In everyday practice, however, scientists understand themselves as being engaged in distinguishing Truth from Error. Correct scientific belief, according to Gilbert and Mulkay, is "thought to follow rather unproblematically from the empirical world" (1982, 167). Accordingly, errors (including "pseudoscientific" ones such as creationism) represent more than oddities or anomalies. Gilbert and Mulkay insist that "most practicing scientists regard the existence of errors as a threat to the enterprise of science itself" (165). Errors, then, become something which must be more than explained; they must be explained away, as outside of or less than science. Consequently, the challenge for evolutionists is somehow to account for the presence (and growing public currency) of the creationist error without doing violence to the working "truths" undergirding and organizing traditional scientific practice.

The discourses of creationism embodied a series of significant challenges to the conceptual integrity of traditional evolutionary science as it had been rhetorically constructed in its rhetorical history. It would be inaccurate, however, to suggest that the evolutionary community was galvanized into concerted rhetorical and practical action by reflection on these interests. Indeed, most practicing scientists regarded the existence of creationism with the kind of irritated and amused incredulity generally reserved for flat earthers (Schadewald 1981-82, 43). It was only when creationism gained public currency and some measure of de facto legitimacy in the broader social context that what had previously been neither a serious demarcation dispute or a political controversy became both.

On this broader level, the issue first implicated in the creationism controversy was perhaps the most fundamental. Creationism had mounted a direct frontal attack on the scientific community's authority to control the public definitions and representations of science and scientific practice. Given the material and political benefits available to science and scientists, the question of who is to decide the constitutive nature of those phenomena is of considerable importance. Bohme and Stehr (1986) have argued that a crucial factor in the maintenance of the epistemically privileged status of science has been its structural ability to control the culturally legitimate definition of what it means to "do science," what it means to justify particular knowledge claims as scientific (Moore 1983, 8–9). This positions demarcation questions near the center of this controversy in particular and the social functions of science in general. The rhetoric of demarcation, then, is not simply an academic exercise; it is a cultural discourse with important social and political implications.

The pivotal site of struggle at which this debate over the "soul of science" was contested was, of course, the public schools (Bliss 1983; Gish 1978, 1983; Gish, Bliss, and Bird 1983; May 1984; Nelkin 1982). The public education system is the intellectual lifeline of contemporary science, because it is the vehicle through which the scientific community can attract new initiates and socialize its membership. This structural link between science and public education alone problematizes the very idea of an insular understanding of demarcation. Demarcation is not the transmission of ahistoric, universal standards. It requires the negotiation of professional, pedagogical, and explicitly political constraints. It is in that process of negotiation that "science" is rhetorically (re)constructed through the relative validation of particular configurations of the ecosystem.

The creationist threat to the prevailing cultural understandings of science was made more significant by the expectations that typical science curricula produce regarding the nature of science. Those expectations served as yet another rhetorical exigence which the evolutionist response would confront. For example, the nature of science textbooks constructs an understanding of scientific knowledge claims as timeless truths, neglecting consideration of the situated contexts in which those claims were manufactured (Knorr-Cetina 1981a). Raths has argued that "in the process of textbook simplification, findings become explanations, explanations become axioms, and tentative judgments may become definitive conclusions" (1973, 211; see Skoog 1983). Such a pedagogical construction of science is more conceptually consistent with the Baconian demarcation advanced by creationist advocates, as well as with the "empiricist folk epistemology" (Cavanaugh 1985) that generally holds sway in public.

More to the point, this construction appears argumentatively vulnerable to the common creationist tactic of pointing out anomalous evidence "against" evolution, such as the incomplete fossil record. The presence of "scientific" evidence superficially at odds with the allegedly secure truths of evolution could serve to "level" the two discourses artificially. As a consequence, creationist aspirations of equal time or balanced treatment in the classroom were legitimated indirectly.

A second implication of the central influence of public schools in this controversy was the threat to the professional interest in the control of curricular issues. Brush insisted that "science teachers have an obligation to help preserve the right to base their teaching on what the vast majority of their professional colleagues accept as science" (1981, 29). The authority for demarcating science, then, is to be found in the scientific community itself. Clearly, the creationist strategy would remove such a right from the traditional community and lodge it in local school boards, state legislatures, or textbook review committees. Recall that despite judicial rejection, several textbook publishers have indirectly altered their discussions of evolution, in apparent deference to creationists' grassroots successes.

Eldredge argued vehemently that "today we have legislatures and judges determining what is and is not science in a manner about as remote from any recognizable intellectual activity as one could imagine" (1982, 149). Whatever the ultimate desirability of such a state of affairs, it entails a view of demarcation as a rhetorical process in which the competing demands and interests of particular sets of social actors and practices are negotiated. The rhetorical choices made and operationalized in those negotiations hold considerable social implications.

It was this nexus of cultural practices that creationism threatened to disrupt and that gave rise to the rhetorical response to creationism. A closer analysis of the rhetorical resources deployed in order to define what it means to "do science" can lead us to a more reflective understanding of the rhetorical "materiality" of science itself.

"And It Is Us": The Response to Creationism Evolves

Perhaps the most striking dimension of the evolutionists' response to the creationist threat was its explicit formulation of the maxim "Science is as science does" (*McLean vs. Arkansas* 1982, sec. IV E). In short, science (in this case, traditional evolutionary science) was articulated, a priori, as the proper and necessary arbiter of all claims and claimants with "scientific" aspirations. The scientific community, then, was defined as growing directly from a normative commitment to intellectual communism (Merton 1973), in which the concern was not so much the evaluation of particular knowledge claims as the broader communal judgment of which claims and/or claimants merited membership in the community. An editorial in *Physics Today* argued, for example, that the community of professional scientists has always and must always be the final arbiter of claims that impact on science or the practices of scientists ("Mainstream Scientists" 1982). The president of the National Association of Biology Teachers emphasized the practical importance of this particular demarcation discourse when he argued, "Given their pretensions of scientific validity for creationism, it seems only proper that . . . they should stand or fall on the scientific merits of their case" (Moyer 1980, 4).

Of course, that which would be taken as "scientific" would follow directly from traditional evolutionary precepts and contemporary understandings of the nature of science. The creationist doctrine, resonating with a older (and justifiably discredited) demarcationist tradition, was automatically excluded from communal membership because it was not "what science did." Moyer continued, "We maintain that creationism is a religious doctrine that cannot be examined within a scientific framework. We cannot verify the existence of a deity or examine our relationship to him using the methods of science" (1980, 4). Viewed in this way, if one claims to be scientific, one ought to be bound by prevailing evaluative standards of science in crafting a response to the encroachment of creationism. After all, the evidentiary support for evolution is overwhelming, and as Prelli noted, "consistency with received knowledge . . . is among the standard presumptions that guide the rhetoric of science" (1990, 320). In this case, however, the rhetorical linkages constituting the ecosystem extended well beyond the daily productive and evaluative activities of professional scientists. Neglecting this interconnectedness, the traditional scientific community's response answered the constitutive question "What is science?" with the normative declaration "Science knows best."

Integral to this response was its central positioning of the assumptions of science and the marginalizing of other evaluative frameworks that might conceivably be relevant to public decision makers. Eldredge, with characteristic vigor, argued that "we must insist on the integrity of our children's education. For scientific illiteracy will send the United States on a straighter and surer path to hell than ever will the idea that we call evolution" (1982, 149). In this way, the community of professional scientists was constructed as having proper authority not only over its own professional concerns but also over broader sociopolitical issues which, by definition, extend beyond the strict confines of science.

In light of Eldredge's credentials and stake in this controversy, one might prefer to overlook his apparent hyperbole. Like much of the response to creationism, however, this sort of response contributed to an artificially rigid distinction between "science" and the broader social contexts in which it is embedded. The social meaning of "science" as social practices is negotiated *within* precisely those social contexts over which science purports to claim authority. While the ultimate reality or desirability of such claims of social authority is not the focus of this study, as a rhetorical demarcation strategy it is a counterproductive method of responding to the creationist threat.

Having maintained that constitutive questions of "scientificness" were said to lie with the community of science, it was left to articulate the relevant standards by which that community would expose the "pseudoscientific" folly of creationism. In his analysis of the creationism controversy, Lessl (1988) maintained that, in the face of the creationist heterodoxy, the traditional scientific community "retreated into orthodoxy"; that is, it rather unreflectively invoked "threadbare epistemic chestnuts" to redefine itself. While there are troubling implications to Lessl's heterodoxy/orthodoxy metaphor (e.g., assuming creationists are part of the *same* science), his analysis is insightful. It suggests that the traditional scientific community drew upon its historically constituted pool of rhetorical resources to reify, rather than reconstruct, its image in the face of a new rhetorical challenge. A closer analysis of the evolutionists' response reveals the deeply contextualized nature of those rhetorical resources and the rhetorical limits and limitations of their deployment.

The most crucial demarcation criterion deployed in the creationism controversy by mainstream scientists was Popper's falsification standard. Popper (1957, 1959, 1962, 1968) maintained that scientific theories could never be verified; instead, the measure of their "scientificness" was said to lie in their relative capacity for potential falsification, that is, the capacity to produce testable predictions which could be proved false by empirical investigation. Those theories not, in principle, capable of disproof were said to be excluded from the domain of the scientific.

What makes the deployment of this standard particularly interesting is that at one point Popper himself had claimed that Darwinian evolution was itself unfalsifiable and hence was "a metaphysical research program" rather than a scientific theory per se (1988, 147). Not surprisingly, creationists seized on this seeming inconsistency, arguing that evolution was, at best, no more scientific than creationism (e.g., Gish 1978, 1988; Morris 1980). Popper, motivated by this creationist maneuver, subsequently "clarified" his position, suggesting that Darwinism was scientific to the extent that "its claims could be falsified through the examination of retrodictive predictions" (Lessl 1988, 27; Ruse 1988).

The visibly absolutist grounding for creationist claims led evolutionists sternly to invoke the falsification standard to exclude them from the domain of the scientific. *Physics Today*, for example, insisted that "the matter is not science. It doesn't represent an experiment which can be falsified" ("Mainstream Scientists" 1982, 54; see Ruse 1988). Similarly, Albert argued, "By definition, 'scientific' creationism is irrevocably grounded in an appeal to the existence and operation of an obviously omnipotent supernatural being. . . . It is absolutely immune from falsification. Literally any problem . . ., can be resolved through an appeal to . . . unknowable supernatural operations" (1986, 27).

What is crucial here is that Popper's formulation was drawn upon in diametrically opposed contexts, to demean the scientific aspirations of both creationism and evolution. This suggests that, rather than a timeless and self-sufficient standard for the demarcation of science, Popper's falsifiability criterion is best conceptualized as a (particularly persuasive) rhetorical resource which can be (and was) deployed to advance particular contextual interests. For the evolutionists, those interests involved the exclusion of a doctrine which, if more widely accepted, would threaten their cultural epistemic authority and the working truths around which their research activities are organized.

While it does seem rather transparently true that creationism cannot be adequately defended as falsifiable, this does not necessarily mean that the evolutionists' deployment of the criterion to discredit creationism is without troubling rhetorical implications of its own. The adequacy of Popper's standard to demarcate science from other social practices has already come under withering philosophical fire (Fuller 1988; L. Laudan 1988b). In no way should falsificationism be understood as unequivocally accepted as *the* conclusive criterion of those activities that we have come to call science. Indeed, a majority of practicing scientists themselves give little, if any, thought to Popper's relevance to their research activities (Potter 1984). With such doubtful status, falsification is insufficient to warrant the definitive implications of its deployment claimed in the response to creationism.

Ironically, arguing that creationism is not science because it is not falsifiable indirectly affirms the creationist call for "equal time." While Popper soon recanted, creationist rhetors were quick to exploit the inconsistency for their own ends. Equivocating on questions of the origin and subsequent development of species, Morris gloated that the shared nonfalsifiability of creationism and evolution meant that "creation at the very least has as much plausibility as a scientific model of origins . . . [and] should be taught in the public schools on at least an equal basis with evolution" (1980, 22). From this perspective, invoking Popperian philosophy to rebut creationism implicitly leveled creationist and evolutionary discourses and empowered calls for balanced treatment.

A second troubling implication of the rather ahistoric invocation and articulation of Popperian falsification relates to the previously discussed assumption of scientific authority over other social practices and discourses. So-called constitutive assumptions of science, when deployed in order to preempt controversy, are objectified and removed from critical study as they are extrapolated into those social practices and discourses. As Aronowitz has argued, "The notions of falsification . . . and other bromides of scientific culture constitute a normative order that can be traced to specific historical roots" (1988b, 533). By presumptively shielding those historical roots from public scrutiny as the necessary locus of decision making broadens to embrace new series of cultural practices, the evolutionist response to creationism reified its own status. This reification added to the mystification which has historically constrained an adequate and humane understanding of the epistemic authority of modern science (Aronowitz 1988b; Mauskopf 1980). Having advanced falsifiability as the fundamental requirement of truly scientific claims and theories, the traditional scientific response constructed a symbolic world in which the creationist discourse no longer threatened the constitutive assumptions of scientific rationality which undergird scientific practice. Simultaneously rearticulating "science" and "accounting" for the creationist error served to situate the cause of that error as "outside the agency of science" (Mulkay and Gilbert 1982a).

The rhetorical deployment of demarcation criteria to achieve this goal first took the form of constructing "disinterestedness" (Merton 1973, 275–77) as a constitutive element of science. In general terms, the communal conduct of science assumes that scientists remain unprejudiced in their interpretation of data bearing on experimental claims. Violations of this norm would then classify the violators as nonscientific. The locus of profound errors of creationists' knowledge claims would subsequently lie outside the domain of the truly scientific, and they would no longer threaten the pristine constitutive rationality of science.

Not surprisingly, the primary cause of "nonscientific" bias or prejudice was articulated as the undeniable (and, to be fair, undenied) religious orientation of the creationists (e.g., Gorman 1981). While there are numerous cases of creationist political advocates self-consciously denying religious motivations in favor, for example, of "fairness" (*McLean vs. Arkansas* 1982, sec. III; Bliss 1983), creation scientists generally tend toward candor regarding the ground of their beliefs (Gish 1978; Morris 1980). A *Nature* editorialist argued that "tolerance has gone too far. It is one thing to be tolerant of religious convictions and quite another to give religiously inspired hypotheses equal time with theories which are scientific in the sense that they purport to relate to the effects and causes which are themselves capable of being understood"

("Does Creation Deserve" 1981, 271). Gould emphasized the incommensurability of creationist convictions and scientific practice when he suggested that "I cannot imagine what potential data could lead creationists to abandon their views. Unbeatable religious belief systems are dogma, *not science*" (1981, 35). Donald Price concluded that "it becomes clear that creationism is simply one more branch of evangelistic apologetics sharing the same goal of preparing the ground for faith and conversion" (1984, 19).

It is important to note here that religious motivations and dogmatism, while not quite construed as equivalent, were closely associated by the implication that such dogmatic motivations would necessarily prevent revision of erroneous beliefs. True scientific beliefs, on the other hand, are constructed as always open to revision in the light of new or more convincing empirical evidence. It was argued, quite rightly I think, that creationist researchers seek to revise interpretations of particular data rather than their fundamental postulates, which are grounded in literalist biblical interpretation.

The rhetorical focus on the religious biases of creationist knowledge claims served to reclaim the pristine working rationality of science. This demarcation rhetoric, however, also had troubling implications of its own. While, as a constitutive question, the epistemological distinctions between science and religion are not always clear, the traditional scientific community's response rhetorically constructed a reductively oppositional and evaluative relationship between them. Ballantine argued that "the suggestion that there is no necessary conflict between science and religion is falsified by the public record" (1982, 11). E. W. MacKie wrote to *Nature*, "You underestimate the menace to science and to the propagation of all rational thought presented by the tolerance of creationism. There is an unbridgeable struggle between science and all forms of mythology" (1981, 403).

While such proclamations were occasionally criticized (e.g., Gilkey 1985; Lippard 1991–92; K. Miller 1984), the rhetorical implications of such an "epistemic hierarchy" are telling. It can easily be read as privileging a particular mode of rationality (i.e., the scientific) over and against other forms of rationality which might be constitutive of other sets of social practices (e.g., religious) which occupy their own position in the social order. Indeed, Nelkin has argued that "scientists hold up for religious discourses the requirement for a direct realism, a literal veridicality, even though they may recognize that this is impossible for science itself" (1982, 41). To this extent, the demarcation of science and scientific rationality constructed in the response to creationism artificially distanced itself from and denied the legitimacy of other social discourses in which rationality is not and need not be defined so narrowly (Boulding 1984).

This exclusion of a particular form of rationality on a priori grounds, while serving to redeem the rationality of science, created an additional rhetorical dilemma. Insofar as science was demarcated as properly unbiased and nondogmatic, the cursory exclusion of creationism (on scientific grounds) from that culture represented a potential inconsistency. Krohn has argued that "science is an open system intellectually, deliberately seeking innovation and accommodating basic change. Yet any system has to work towards intellectual coherence or it would fall into anarchy. This has never been programmatically resolved in science" (1977, 95). This study would suggest that such a tension between coherence and innovation, what Kuhn (1977) called the essential tension, cannot be resolved programmatically, but must be managed in particular rhetorical encounters.

An additional demarcation strategy invoked in the attempt to manage this tension involved the identification of creationist advocates as incompetent to be granted the label of scientist. Mulkay and Gilbert (1982a) argued that a common strategy in scientific controversy in which the rationality of science is at stake is simply to claim that the offending party simply doesn't understand the relevant issue. As such, the *competent* scientist, which Mulkay (1975) suggested was the only true scientist, is one who has a functional grasp of what is taken to be the relevant information. There was very little dispute over the legitimacy of "science" as a decision rule, so it was left to evolutionists to construct a narrative in which the incompetence of creationists explained their erroneous claims.

A notable example of this strategy involved the (mis)use of the term *theory* throughout the creationist discourse. The intuitively appealing, though woefully imprecise, notion of theory as little more than an educated guess or formulation of opinion is endemic to the creationist doctrine. As Gish put it, "Evolution is *just* a theory, as is creation. There are no justifiable grounds for excluding only one of them from the classroom" (Gish 1978, 136; Gish, Bliss, and Bird 1983). Just as obviously, the notion of theory which is operationalized in traditional scientific practice is that of a "formulation of basic principles in a particular topical domain which is supported by empirical evidence and open to confirmation or falsification by evidence yet to be discovered" (Gilkey 1985, 127–37; Root-Bernstein 1984b). Kitcher concluded that we should "reject the creationists' gambit. . . . Science is not a body of demonstrated truths. Virtually all of science is an exercise in believing what we cannot *definitively* prove" (1982, 32).

Of course, such a bald admission of the indeterminacy of evolution actively militated against its hallowed epistemic status. In what can be read as an attempt to negotiate these competing rhetorical demands, the evolutionists' response drew a meaningful distinction between the more epistemically secure *fact* of the existence of evolutionary process and the less secure knowledge of the mechanism by which evolution is actually accomplished. The creationists' failure to grasp this distinction was said to be compelling evidence of their sheer incompetence. Lewontin, for example, argued that "it is time for students to state clearly that . . . what is at issue in biology are questions of the details of the process and the relative importance of different mechanisms of evolution" (1981, 559). Root-Bernstein suggested that this phenomena/mechanism issue was the most important "point of confusion in the evolutionist vs. creationist debate" (1984b, 71). As the evolutionist response constructed it, the creationists' failure or inability to recognize the theoretical and practical consequences of this distinction lay at the base of their scientific incompetence.

A second instance of the rhetorical construction of "incompetence" involved the creationists' consistent (mis)interpretation of the Second Law of Thermodynamics. This physical law states that, over time, the total supply of energy tends to run down and move in the direction of increased entropy (V. Anderson 1983; Freske 1983; Patterson 1983). Creationists argue frequently that evolution, with its predicted progression from less ordered to more highly ordered life-forms, represents a particularly egregious violation of the law. Gish argues that "it is difficult to understand how a discerning person could fail to see the basic contradiction between these two processes. It seems apparent that both cannot be true, but no modern scientist would dare challenge the validity of the Second Law of Thermodynamics" (1983, 185). The creationist alternative is to challenge, and reject, evolution instead.

The traditional scientific community's response was definitive, portraying creationist arguments as predicated on the false assumption that the biosphere is a closed system in which the Second Law is known to be relevant. Of course, the biosphere constantly receives energy, in the form of sunlight. Cracraft argued that "numerous recent studies have successfully applied thermodynamics, properly understood, directly to the question of biological evolution" (1983, 112), suggesting that the issue of the Second Law did not hinge on matters of interpretation of equivocal data. It turned, instead, on the fundamental misunderstanding of an unequivocal theoretical postulate. Patterson offered two conclusions regarding the creationist interpretation of the Second Law: "that the creationists' second law arguments are seriously mistaken, and . . . that the mistakes reflect principles and concepts covered in beginning level courses in thermodynamics" (1984, 134). The creationists' error, then, was a function of their own incompetence and hence could not indict truly scientific assumptions of rationality.

The final rhetorical obstacle that organized the response to creationism was the curious position of creation scientists as simultaneously within and yet excluded from the community of professional scientists. Recall that the majority of creationist advocates had been granted advanced degrees in scientific (or technical) disciplines from generally respected universities. Hence, the above attributions of error indirectly served to challenge the legitimacy of the system of quality control within the community of science.

The rhetorical response to creationism gave evidence of just such a dilemma. A common strategy was to suggest that creationists no longer bore the imprimatur of science because they had willfully violated the norm of "intellectual communism" (Merton 1973, 273–75). Insofar as they had very self-consciously sought out public approbation via school board policies or legislative mandates, creationists had betrayed the ethos of the scientific community. Creationists, it was argued, had violated the constitutive normative structure of science by extending the platform of debate to include the general public via such institutions as legislatures and school boards.

Accordingly, the scientific community's response to creationism sharply criticized the public advocacy activities of the creationists. Derisive commentary abounded in relation to the "merely rhetorical" victories achieved by creationists armed with the *Handy Dandy Evolution Refuter* in public debates with "well-intentioned, but unprepared evolutionists" (Lewin 1981, 635). Kenneth Miller noted that "such debates, of course, are neither part of science nor a contribution of anything to scientific understanding. Their purpose is political" (1983, 249). Eldredge indignantly argued that "the [creationist] literature is all written in extremely simple terms, to convince the uninitiated rather than to enter into a sophisticated scientific dialogue" (1982, 83). According to Beyerstein, "Scientists . . . must not only persuade their audience, they must also present the scientific facts fairly according to the canons of scientific discourse. . . . Creationists . . . labor under no such constraint" (1990, 20).

While creationists often simplify to the point of misleading trivialization, the traditional scientific community's response went well beyond this judgment to a portrayal of science as functionally and conceptually immune from the broader social contexts in which it is embedded and through which it is partially defined. This insular demarcation of science reifies the detachment of science from that context while simultaneously affirming its epistemic authority over it, perpetuating the mystification which leads to public unease with the perfectly reasonable public policy contributions of science (Culliton 1978, 147–55).

The foregoing analysis has offered tentative evidence for the heuristic utility of conceptualizing the demarcation of science as rhetorical practices emerging from specific sets of social conditions and as responses to challenges to particular cognitive and professional interests. In this case study, I have argued that, out of a wide variety of potential demarcation criteria, evolutionists deployed certain of them to construct a working identity of science which excluded creationism and its erroneous knowledge claims. This rhetorical demarcation had significant political and epistemic consequences. It served to reify the primacy of "science" over other systems of social practice, defined its proper role as arbiter of contested public epistemologies, and actively distanced itself from the constitutive practices of the social contexts of which it was a part. In the final section of this chapter I will briefly speculate on the rhetorical limits and limitations of this demarcation in its implications for the relationship of science and society.

The Social Costs of Demarcation

By failing to recognize its own limitations, the demarcation of science accomplished in response to the emergence of creationism was, in many ways, a greater threat to science than creationism could ever hope to be. While demonstrating the empirical weakness of the creationist doctrine, it also constructed a "science" that artificially and counterproductively distanced itself from the broader social contexts that actively constrain the nature and functions of contemporary science. Its denigration of alternative modes of rationality and its unreflective assumption of authority over traditionally social institutions only fan the flames of anti-intellectualism that spawn such reactionary movements as creationism. What is called for, then, is a new construction, one that promotes the fruitful conduct and rigorous evaluation of scientific research, yet is able to interrogate and modify its relationships with other, different, yet equally valuable sets of social practice.

Tolstoy once observed that "science is meaningless because it gives us no answer to the question, the only question important for us: what shall we do and how shall we live" (in Weber 1958, 143). Tolstoy's hyperbole raises an issue central to this discussion: In what ways can science and other technical discourses contribute legitimately to deliberation in the public realm? Put another way, how can technical discourses contribute to our ongoing search for *phronesis* in an age when technical expertise and practical wisdom must somehow coexist? (Willard 1990, 145).

The tensions between technical and practical wisdom, and hence the demarcations enabling them, are magnified when the former is portrayed as contributing to the erosion of traditional bastions of authority and morality. A National Science Foundation study demonstrated that, between 1957 and 1979, the percentage of Americans who believed such a erosion was occurring rose from 11 percent to 27 percent among those who considered themselves attentive to science. Among the general public, that increase was even more compelling: from 24 to 42 percent (Miller, Prewitt, and Pearson 1980; Yankelovich 1982). It is precisely this ambivalence that frames and makes problematic the demarcation practices wherein the community of "science" is socially constructed for its own members as well as for consumers of its knowledge products (Gieryn 1983b; Holmquest 1990; Taylor 1991a).

The existence of more than negligible ambivalence toward science in general and evolution in particular suggests that the demarcation of science considered here succeeded in maintaining its "uniqueness" among social and intellectual activities (Mulkay 1975). That success, however, might have been purchased at the high cost of a measure of its own public authority. The imperfect fit between the domains demarcated in this case illustrates the rhetorical limitations of technical expertise as it is articulated before nontechnical audiences.

Initially, the vision of science constructed in the response to creationism was highly insular, holding that the conduct of scientific business, in any or all of its forms, including education, is an activity best organized and controlled by technical experts. Whatever its intrinsic merit, such a perspective is clearly at odds with the prevailing populist conception of public education. Nelkin has argued that public school education is among the most volatile areas of public policy (1982, 55–70, 165–84). As a result, the realities of interest group politics will, for better or worse, influence the nature and content of curricular material.

The equivocation of technical expertise regarding evolution with the nontechnical and inescapably political authority to control public education promoted the growth of an oppositional discourse with troubling implications for the quality of science *and* public education. The nagging persistence of creationism and its related call for increased control over science textbooks is a reaction to the perceived threat to local control and authority represented by a feared scientific hegemony. Indeed, Feyerabend, calling attention to the implicit totalitarianism of modern science (with tongue only partially in cheek), remarked, "Three cheers go to the creationists . . . but I know that they would become as chauvinistic as scientists are today when given the chance to run society all by themselves. . . . Ideologies become doctrinaire as soon as their merits lead to removal of their opponents" (in Brush 1986, 169). As a rhetorical practice, bound at some level to its consequences, assuming authority appears far less compelling and more problematic than arguing for it.

A second troubling implication of this contextualized demarcation involves the contested social and epistemological turf claimed by "science" and "religion." While most scientists may not, as a matter of principle, reify a dichotomy between scientific rationality and religious irrationality, this rhetorical process of demarcation constructed just such a division. Gilkey, reflecting on his experiences in the Arkansas creationism trial, argued that "the assumption that religion represents early myths now outmoded by scientific developments pervades much of the writing and teaching of science" (1985, 113). He decried the assumption "that as scientific knowledge accumulates, the ignorance represented by religious myth will be dissipated" (180). As such, the community of science constructed here was predicated on the subordination, if not elimination, of arguably less rational modes of cognition, such as religion.

While the ultimate reality of such a division is beyond the scope of this discussion, the rhetorical impact of such a division merits attention. It cannot be disputed that, among sets of social practices, "the religious is a permanent, pervasive, and always central aspect of human life, generated out of fundamental human capacities and needs" (Graham 1978, 5). Geertz (1968, 1973), for example, argued that the heart of the religious perspective has little to do with a heaven or an afterlife, but rather with the conviction that the values that one holds are grounded in an inherent structure of reality. Many people are likely to view religious belief, rather than science, as best suited to answer fundamental questions of value.

By denigrating such an important and pervasive dimension of social life (implicitly or otherwise), the evolutionist response represented itself as unresponsive, even alien, to the needs (rational or otherwise) of the broader culture. That culture has been constituted with a number of ways of knowing, only a very few of which have ever been construed as "scientific." Hence, the narrow rhetorical circumscription of the legitimate modes of rationality simultaneously circumscribed the cultural relevance to which science could lay claim.

As critics, we must affirm the legitimacy of technical expertise in those deliberations which require it. The rationality of public deliberation is based, in no small measure, on the judicious selection of relevant experts and expertise. At the same time, however, we must avoid granting authority to expertise simply because it is expert (Derber, Schwartz, and Magrass 1990; Feyerabend 1978). To do so is to usher in what Mikhail Bakunin called "the reign of scientific intelligence, the most aristocratic, despotic, arrogant, and elitist of all regimes" (in Chomsky 1982, 61). The public demarcation rhetorics of the scientific community demand special scrutiny in this regard, because nontechnical audiences may craft oppositional discourses to deal with what they perceive as the unwarranted extension of expertise in the domain of public authority.

Blame, if it is to be assigned at all, must lie fundamentally with the discourses of expertise which claim too much for themselves and allow too little from nonexpert audiences. As Wynne concluded, "Scientific . . . institutions that want to integrate science into lay public lives must be organized so as better to understand . . . public agendas . . . rather [than] . . . to impose a scientific framework of understanding as if that on its own were adequate" (1991, 115). Perhaps a deeper appreciation for those nontechnical audiences and more critical attention to the rhetorical constructions of expertise and authority will produce clearer understanding of their inescapable symbiosis.

6 DEMARCATING COLD FUSION

RATIONALITY, MATERIAL INTERESTS, AND "GOOD SCIENCE"

On 23 March 1989, University of Utah chemist B. Stanley Pons and his mentor from the University of Southampton, Martin Fleischmann, announced at a Salt Lake City press conference the results of their five-year research program. The headline of their press release read, "Simple experiments result in sustained n-fusion at room temperature for the first time; Breakthrough process has the potential to provide inexhaustible source of energy" (Bishop and Wells 1989; Broad 1989c; Peat 1989). Using equipment that could be found in almost any high school chemistry laboratory, they claimed to have, in effect, harnessed the power of the sun. Their apparatus had apparently produced fusion energy for more than one hundred hours in a laboratory beaker filled with "heavy water," a deuterium-based variation of ordinary seawater (Fleischmann and Pons 1989; Fleischmann, Pons, and Hawkins 1989).

It would be no exaggeration to suggest that the scientific community was thrown into a state of excitement and confusion (Cooke 1989; Cookson 1989; Fitzpatrick and Wilson 1989). Fax machines, computer networks, and electronic bulletin boards began to circulate updates, questions, and criticisms of the Utah experiments. Around the globe, scientists scrambled to replicate the experiments which threatened to render more than four decades and millions of dollars' worth of previous fusion research obsolete in one fell swoop (M. Simons 1989). Compared with the multimillion-dollar high-temperature plasma containment devices or processes of laser bombardment long thought necessary for fusion research, the Utah experiment appeared maddeningly simple. It required only a laboratory flask filled with heavy water and lithium salts, a palladium rod encircled by a platinum coil, and an electrical source. Two weeks after they began to pass an electrical current through the apparatus, Pons and Fleischmann claimed to have measured the output of four watts of energy for one watt of energy input.[1] Their explanation was that the deuterium ions in the heavy water migrated into the crystal lattice of the palladium, where they were held in close enough proximity for long enough periods of time that they overcame their natural repulsion and fused. This was said to produce heat and low levels of radiation, signature products of fusion reactions (Fleischmann, Pons, and Hawkins 1989; Unger 1989; Van 1989a).

As the scientific community struggled to make sense of these startling claims, the federal government moved quickly to investigate their technological and economic implications. On 24 April 1989, James Watkins, secretary of the Department of Energy, noting the "potential benefits from practical fusion energy," requested that the Energy Research Advisory Board (ERAB) convene a special panel to investigate the Utah claims and other cold fusion research (ERAB 1989, 39). On 26 April, the House Committee on Science, Space, and Technology held hearings on the fusion findings and the University of Utah's request for twenty-five million dollars to establish a fusion research center near its campus (*Recent Developments* 1989). Meetings of the Electrochemical Society and the American Physical Society convened to discuss the Utah findings and subsequent research (Pool 1989a, 1989b; L. Young 1989). A special international conference at Los Alamos National Laboratory provided a host of challenges to Pons and Fleischmann's techniques and conclusions. A conference, cosponsored by the Electrical Power Research Institute and the National Science Foundation's Division of Electrical and Communication Systems and featuring a disproportionate number of fusion supporters, was convened.[2]

Ultimately, the furor over cold fusion receded in the face of mounting negative evidence and questions about the laboratory practices and personal motivations of the Utah team (Allegretti 1989; McCann 1989; Rolly 1989). Perhaps the most telling blow was the 26 November 1989 release of the Energy Research Advisory Board's final report, which concluded that no funding should be given to the Utah project and generally dismissed the Utah claims as anomalous (ERAB 1989).

In retrospect, it appears that Pons and Fleischmann had really accomplished little more than to confirm Andy Warhol's hypothesis that everyone would be famous for fifteen minutes. Their professional reputations sullied and their experimental claims lambasted, Pons and

1. It should be noted, however, that this claim at the press conference is not substantiated in the published account of the research (Fleischmann and Pons 1989; Fleischmann, Pons, and Hawkins 1989), less yet by independent researchers.

2. The physics and nuclear physics divisions of the NSF had declined cosponsorship, citing widespread doubts about cold fusion in general and the Utah studies in particular. See Huizenga 1992, 150ff.

Fleischmann have faded from the scientific and public spotlight into the relative obscurity of a French lab, while limited (and far less publicized) fusion research projects continue in other laboratories ("Cold Fusion's Discoverers" 1992). Whatever the final ontological status of cold fusion, this episode of scientific controversy is ripe with rhetorical significance. This chapter focuses on what I take to be one crucial dimension: the rhetorical processes wherein the conceptual and practical boundaries of science were articulated in the response to the Utah claims. In short, this chapter will explicate how demarcation was rhetorically managed in the course of what was generally taken as simply an internal scientific dispute.

Implicit Demarcation Practices in Orthodox Science

While the most common emphasis in discussions of the demarcation of science has been on the exclusion of various "nonsciences" or "pseudosciences," processes of interdisciplinary demarcation are of equal rhetorical and social importance. The fact that relatively little attention has traditionally been given the demarcation dynamics of controversies internal to science does not justify continued neglect. Indeed, I will argue that arbitrary distinctions between those sets of practices lying outside of science and those thought to lie within it are empirically and critically unwarranted.

The cold fusion controversy of 1989 represented one especially appropriate case study in the rhetorical creation and maintenance of multiple goal-oriented lines of demarcation. The response to the highly controversial claims advanced by Pons and Fleischmann constructed a constitutive scientific ethos which functioned rhetorically to sustain important communal interests by excluding particular practices framed as "nonscientific." The operative definitions of science which rhetorically emerged in this discourse are explicable only in relation to the cognitive, technical, and professional interests of rationality and social epistemic privilege which they sustain. Beyond this demarcation between science and nonscience, however, an interdisciplinary demarcation was constructed so as to maintain a particular disciplinary hierarchy. This demarcation rhetoric gives evidence of deeply contextualized rhetorical accommodations between normative communal interests, broader social and political constraints, and practical instrumental goals. These demarcation practices actively constrained the empirical claims advanced and defended by the competitors in this scientific debate.

Many critics, of course, will respond that the cold fusion controversy represents little more than a case in which two scientists working in an area far removed from their own expertise made claims which were wrong (e.g., Chapline 1989) or perhaps even fraudulent (Close 1990). Whether that is in fact the case is of little consequence to this project. The ultimate ontological status of cold fusion processes is a question better left to others. This seems especially true of the early stages of the controversy, in which the ontological questions remained open, whatever their face plausibility. I will argue that it is precisely within the constructions of the "grounds for error" that the boundaries of the scientific enterprise are rhetorically articulated.[3] The presumptions regarding the constitutive nature of science remain utterly tacit during periods of Kuhnian normal science. Logically, then, it is only during periods of scientific controversy that these presumptions are made explicit and deployed to advance particular personal, professional, technical, and/or cognitive interests.

Textual indications of the rhetorical demarcation practices at work in those controversies typically considered to lie comfortably in the realm of "real science" are especially convincing evidence of the socially constructed nature of "real science." Demarcation practices are commonly assumed to be a dominant resource in discussions of creationism, Lysenkoism, or flat earthism. Consequently, it is not particularly difficult to find garden variety demarcation practices in such discussions (although one can certainly find novel demarcation strategies). It is in so-called hard cases, however, that demarcation practices are thought to be irrelevant and where evidence of their formative influences would hence be of special theoretical interest. Latour and Woolgar, for example, attempting to demonstrate the "social construction of scientific fact" in the genesis of the Thyrotropin Releasing Hormone, argued that "if the process of social construction can be demonstrated for a fact of such apparent solidity, . . . this would provide a telling argument" for the ubiquity of those processes with which the strong program was concerned (1979, 106). Similarly, I will argue that especially convincing evidence of the centrality of rhetorical demarcation practices can be found in a case in which they are not self-evidently relevant.

This chapter analyzes the discourses of fusion supporters and critics during the first year of the controversy. My goal is not to provide the ex-

3. Huizenga, while resisting a view of the cold fusion case as representative, observes that "the cold fusion fiasco illustrates once again, as N rays and polywater did earlier, that the scientific process works by exposing and correcting its own errors" (1992, 236). Whatever the ontological status of the Utah errors, I am concerned here with the rhetorical processes of error exposition and correction as a function of demarcation.

haustive and definitive account of the entire cold fusion controversy.[4] For example, the responses of and to Brigham Young University physicist Steven Jones and his muon catalyzed cold fusion model will not be a primary focus of this study but will serve only as a rhetorical foil against which the responses to and of Pons and Fleischmann might occasionally be compared.[5]

The text base considered here consists of five types of discourse: news accounts of the controversy; the transcript of the 26 April hearings before the House Committee on Science, Space, and Technology (*Recent Developments* 1989); the ERAB's final report (1989); the major semi-technical histories of the controversy (Peat 1989; Close 1990; Mallove 1991; Huizenga 1992); and the technical journals in which the competing claims were articulated most precisely. The primary journals included were *Nature*, the *Journal of Electroanalytical Chemistry and Interfacial Electrochemistry*, *Physical Review Letters*, and *Fusion Technology*. In addition, bridge-level journals such as *Chemistry and Industry* were included when articles dealt with technical issues, rather than, say, with strictly economic implications for industry.

I certainly recognize that this sample is highly diverse, and even potentially confusing. However, it was chosen for important theoretical and critical reasons. Initially, we should problematize the unreflective distinctions too often drawn between the internal and external discourses of and about science. Such distinctions are themselves products of a too often naive demarcation which serves to reinforce the unreflective epistemic privilege of modern science. Science, as conceptualized here, is a shifting series of social practices with necessary and symbiotic relationships with other sets of social practices. To label particular discourses, a priori, as intrinsically scientific or nonscientific is to ignore or deny their reciprocal influences.

The newspaper accounts used in this study consisted of all articles related to cold fusion and/or Pons and Fleischmann which appeared between 22 March and 29 December 1989 in the *New York Times*, the *Los*

4. The most complete (though often contradictory) histories of the controversy are Close 1990; and Mallove 1991. Close is harshly critical of the Utah studies and even more so of Pons and Fleischmann's behavior, while Mallove appears to accept the reality of cold fusion and condemns what he perceives as the scientific community's unfair treatment of the two. Another interesting account, this from the chair of the ERAB's Cold Fusion Panel, can be found in Huizenga 1992. The evaluative tone of the volume is foreshadowed in its title: *Cold Fusion: The Scientific Fiasco of the Century*.

5. Ironically, Huizenga, one of cold fusion's staunchest opponents, would extract the BYU research from the *cold fusion* saga altogether. As he describes it, "Although muon catalyzed fusion is often called cold fusion, it is a well understood process" (1992, 15). He reiterates that distinction at least nine times in his book-length account of the controversy.

Angeles Times, the Salt Lake City *Deseret News*, and the *Chicago Tribune*. Additionally, a search of the Newsbank Index yielded 143 articles which appeared in generally smaller dailies across the nation. Analysis was restricted to quoted statements from persons identified as scientists. To the extent that this project is not, in principle, a study of the news coverage of the cold fusion dispute, every effort was made to exclude broader editorial characterizations of the controversy (Lievrouw 1990).

Of course, it could well be argued that the utilization of news accounts is necessarily complicated by the journalistic tendency toward the popularization of science (LaFollette 1990; Nelkin 1987; Taylor and Condit 1988). It is also interesting to note that, at least in the early stages of the controversy, practicing scientists were themselves limited to news accounts for their own information (Peat 1989, 89–90). Stanford chemist Robert Huggins, for example, attempted to reconstruct the Utah apparatus from film footage on public television's *MacNeil-Lehrer Newshour*. In order to gauge its size, he made calculations based on the apparent width of Pons's wrist (Jacobsen-Wells 1989b, 28: E8). An installment of *Nova* (Lemonick 1989), pictured an MIT research team "wearing out a videotape" in attempts to re-create the Utah apparatus, as seen in broadcast news accounts.

Additionally, it remains unclear precisely to what extent any such "popularization" substantively alters the nature of science as social practice. This study problematizes the presumptive barrier between the scientific and the social. Cloitre and Shinn, for example, insist that "the traditional argument that scientific popularization is somehow radically different from other classes of science associated texts . . . tends to collapse" (1985, 58). This is not to say that they all sound alike. Clearly there are important distinctions. Those distinctions, however, are the products of situational rhetorical ends and not of principled epistemic necessity.

The inclusion of the House hearings might also be challenged on the grounds that they represent the intrusion of ancillary political and financial considerations into the realm of the truly scientific. As such, they would simply reflect people talking to politicians *about* science. I contend, however, that the scientists testifying before Congress provided compelling evidence for the symbiosis of demarcation practices with the broader political and financial considerations which organize congressional hearings. Their testimony, along with the other supporting materials, reveals the inherent plasticity and open-endedness of science. As a central element in the "scientific ecosystem" crafted in this controversy, such considerations were far from ancillary to the question of what we take science to be. Indeed, they are formative influences on the definition and practices of contemporary science.

While the ERAB's report, *Cold Fusion Research,* and the sample of professional journals carry the most face validity as scientific discourse, this need not entail that they occupy a privileged interpretive space. Nonetheless, one rhetorical function that such technical literature serves is that of demarcating its own topical domain from other domains. This is not to say that many (or any) journal articles directly take up the question "What is science?" Rather, it is to say that when scientific rhetors write experimental papers, they claim credibility by linking themselves discursively to particular communal assumptions and commitments. By illustrating how scientists craft these linkages, we can understand how the impersonality and/or universality which are generally taken as essential to science are constructed. In this way, the patterns of social accounting in technical journals can be read as the negotiated construction and legitimation of a particular cultural identity. In short, the articles can be read as an exercise in the rhetorical demarcation of science.

Reclaiming Rationality: Demarcation and the Locus of Error

The crucial question at issue in the cold fusion controversy was whether, in fact, the Utah researchers had produced nuclear fusion in their basement laboratory on the Utah campus. Allowing for widespread doubt, it might be put more succinctly as "How did they get it so wrong?" The common criticisms of their experimental activities were generally oriented toward constructing an evidential context in which it became reasonable or unreasonable to accept their experimental claims as valid. One might, then, be tempted to conclude that the cold fusion controversy was fundamentally a question of fact. That is, did they discover it or not?

The rhetorical discourses that were marshaled in an attempt to provide coherent and convincing answers to that question clearly revealed the contextual (re)definition of the community in which those discourses were advanced, that is, the community of science. In the process of evaluating the relative veracity of a purported "scientific fact," competing rhetors actively, albeit perhaps unintentionally, constructed the standards by which any "fact" could potentially be considered "scientific." Whitley has cogently argued that "'facts' are socially constructed cognitive objects, liable to reinterpretation and change, which become established through negotiations . . . among scientists" (1985, 11).

This case study suggests that in the process of adjudicating the communal acceptability of the alleged findings of high-level cold fusion, scientific rhetors characterized the constitutively normative nature of science so as to maintain presumptive conceptions of scientific rationality and to sustain important structural and professional interests. These demarcation rhetorics functioned recursively to constrain the perceptual, interpretive, and evaluative judgments brought to bear on the facticity of Pons and Fleischmann's controversial claims.

The conceptions stem from the traditional assumption that scientific knowledge is somewhat distinct from, indeed superior to, other ways of knowing (Aronowitz 1988a). In his discussion of popularization, Whitley notes that the "conventional view . . . conceives scientific knowledge to be unitary and epistemologically privileged" (1985, 10). I do not raise this issue on ontological grounds. The ultimate truth of cold fusion (or any other scientific claim) does not concern me here. I want only to characterize this conception of rationality as a crucial constituent of the social configuration that grants an overwhelming epistemic privilege to knowledge claims bearing the imprimatur of science.

While disagreement is the lifeblood of science, too much of a good thing isn't always a good thing. The working assumption of the pristine rationality of "truly" scientific claims serves as a dominant professional and cognitive interest that is threatened by periods of sustained scientific controversy regarding issues long thought settled. While many scientists might reject a conception of scientific truth as some sort of ontologically pure state, such assumptions are constitutive constraints on scientific practice, that is, on the everyday productive and evaluative actions of scientists as laboratory laborers.

Scientists, in everyday practice, understand themselves as distinguishing truth from error (Potter 1984; Shapin 1980; Ullrich 1979). Correct or true belief is thought to follow rather unproblematically from the empirical world, which, of course, is understood as uniform and stable, albeit mysteriously so. Hence, factual errors, particularly profound ones, represent more than intriguing experimental anomalies or professional annoyances. Gilbert and Mulkay (1982) suggest that most practicing scientists regard the existence of errors as fundamentally threatening to the conduct of science itself.

This is not to say that science and scientists somehow hide error. Indeed, it and they might properly be understood as seeking out and accounting for errors more rigorously than in other social and intellectual activities. My concern is with the professional rhetorical response to such errors, once they are discovered. For in those moments when the received wisdom of particular research specialties (e.g., paleontology or fusion physics) is called into question, the phenomenal world is made to appear obscure. In moments of resolution, when an account for that error can be sustained, however, that world becomes again more orderly. The earlier offending disorder and recalcitrance are then displayed as defects in the *human* contribution to knowledge making. Errors, then, must be explained as the product of nonscientific influences.

This implicit objectivism subtly constrained the rhetors in the cold fusion controversy. Fleischmann and Pons, for example, in discussing the natural properties of hydrogen and deuterium, contended unequivocally that "in view of the very high compression and mobility of the dissolved species *there must* therefore be a significant number of close collisions" (1989, 302; emphasis added). Notice that the characteristics of natural elements were articulated as necessarily orderly reflections of an empirical world of equal order and regularity. Articulating an identical conception of the natural world for diametrically opposed purposes, two physicists very critical of the Utah announcements utterly rejected claims of the suppression of the coulomb barrier which must precede fusion activity (Leggett and Baym 1989). Nonetheless, they defended their criticism on grounds that presume the objectivity of the physical world and its *truly scientific* investigation, arguing that "unless the [Utah observations are] quite anomalous, the Coulomb barrier penetration in a solid in equilibrium *cannot* be enhanced anywhere near the magnitude required to explain the fusion rates inferred from the experiments" (Leggett and Baym 1989, 45; emphasis added). While alternative contingencies (e.g., anomalies) are mentioned here, it is important to note that they are characterized as inexplicable, hence not properly supported. This failing was attributed to the explanatory account, rather than to the reality it had failed to explain. For practical purposes, modal qualification was excluded from descriptions of the natural world and its implications for the potential facticity of the alleged cold fusion processes. In our terms, then, what is important is that science is implicitly demarcated as the rational explication of an equally rational natural world.

It was equally apparent that the advancement (not to mention unparalleled publicizing) of the controversial claims of cold fusion represented a significant threat to the view of the world that had long underwritten fusion research (Huizenga 1992; Peat 1989, 20–63; Close 1990, 83–105). The problem of producing, maintaining, and containing a fusion reaction had been a central concern of research in physics for several decades. While techniques and success rates had varied considerably, the goal of all such experiments—room temperature or otherwise—had always been to cause hydrogen atoms to overcome their natural electrical repulsion and join, creating helium atoms and releasing large amounts of energy. Inducing fusion to occur, however, has proved an elusive accomplishment (Cooke 1989, 41: E1). Since two deuterium nuclei possess identical electrical charges, they naturally repel each other. Although there are some important exceptions to the rule (e.g., S. Jones et al. 1989; Rafelski and Jones 1987), until Fleischmann, Pons, and Hawkins' controversial paper was published (1989), conventional scientific wisdom held that the only plausible method of overcoming this repulsion was to force the nuclei to collide at such high speeds and temperatures that the natural repulsion of the coulomb barrier would be breached and fusion could take place.

In pursuit of such controlled fusion, physicists at major fusion laboratories such as MIT, Princeton, and CalTech had long employed multimillion-dollar research facilities in their efforts to enact those experimental conditions. Perhaps the dominant North American approach involves high-temperature plasma containment devices designed to facilitate the extremely high temperatures required to produce the high rates of collision speed required for fusion reactions (Herman 1990, 82ff.; Peat 1989, 34–37). In these huge "tokamak" reactors, the superheated plasma is confined in extremely powerful magnetic fields. Such magnetic suspension is required because no known solid material is capable of containing plasma at temperatures of up to one hundred million degrees (Miley, in *Recent Developments* 1989, 131–35).

A second approach to fusion research involves equally expensive equipment and processes of "inertial confinement," which involve bombarding a deuterium and tritium pellet with powerful laser, muon, or other particle beams. Here the intent is not to achieve high temperatures but rather to squeeze the hydrogen atoms together so tightly that they overcome their natural repulsion and fuse, liberating energy. Additional technical details are available elsewhere (e.g., Detjen 1989a; ERAB 1989; Furth, in "*Recent Developments* 1989, 168–77, 105–13, 127–45, 149–67; D. Williams et al. 1989).

Here, I want only to indicate that the 23 March announcement of the Utah findings represented a significant threat to the professionally sustained interest in scientific rationality. Contrary to the "received wisdom" of fusion physics and the huge government grants which produced it, two chemists now publicly claimed to have discovered room-temperature fusion on the cheap. As a consequence, the claims of Pons and Fleischmann represented a profound puzzle, a challenge to the received view of the physical world that had underwritten fusion physics and hence functioned as the standard of rationality to which any knowledge claim had been held accountable. As Huizenga put it, "Enhancing the probability of a nuclear reaction by 50 orders of magnitude . . . via the chemical environment of a metallic lattice, contradicted the very foundation of nuclear science" (1992, viii). MIT fusion physicist Ians Hutchinson provided a more humorous description of the potential impact of the Utah claims: "Suppose you were designing jet airplanes and then you suddenly heard on CBS News that somebody had invented an anti-gravity machine. . . . That's the way I feel" (in Detjen 1989b, 40: D9). In short, one was led to wonder how science had gotten it so wrong for so long.

If modern science is constrained ideologically to demonstrate its own objectivity and, by implication, the regularity of its object of investigation, then the findings announced by Pons and Fleischmann in Salt Lake City posed a significant rhetorical dilemma. The traditional fusion community either had to reject the accumulated results of some forty-five years of painstaking (and expensive) research or had to account somehow for the obscure and recalcitrant Utah findings in a manner which reclaimed for scientific practice its ostensible grounding in the terra firma of objectivity.

It was this rhetorical situation in which the scientific community found itself in the days and weeks following the Utah press conference. The rhetorical task was a complex one: to give the "heterodox" fusion claims a "fair hearing" while preserving the conception of rationality that underwrites everyday scientific practice and had underwritten decades of fusion research. The rehabilitation of scientific rationality was a key constituent of the rhetoric advanced in the media, in technical journals, and before Congress.

A Constitutively Normative Account of Error

Like the rhetorical response to "scientific creationism," the response to cold fusion was dominated by its advancement of a constitutive ethos for the scientific community. The construction of such an ethical framework for understanding what it means to "do science" did not itself bear directly on the ontological facticity of cold fusion. Rather, its substantive rhetorical function was to demarcate the domain of the truly scientific in such a way as to situate what were represented as the *causes* of the Utah team's experimental error *outside* that domain.

Without adopting its epideictic dimension, Merton's (1973) pioneering analysis of the normative structure of the scientific community offers a useful vocabulary for mapping the rhetorical dynamics of this controversy. In the process of that mapping, however, the limits of traditional vocabularies such as Merton's are illuminated as well. The rhetorical deployment of the norms of disinterestedness, communality, and organized skepticism constructed a constitutive ethos of science which effectively served both to delegitimate the discourse of cold fusion and to reclaim the rationality undergirding previous fusion research practices.

In light of the skepticism circulating throughout the traditional fusion research community, it is perhaps wise to consider first the rhetorical enactment of what Merton called the normative commitment to organized skepticism. In practical terms, the scientific community is said to be committed normatively to the suspension of belief until adequate evidence is available to warrant acceptance of the novel knowledge claim in question (which, of course, is a function of the operation of peer review, full disclosure, replication, and so on). Claims that Pons and Fleischmann had unjustifiably breached this normative prescription were common. As the dominant interpretation of the cold fusion claims had it, fools had indeed rushed in where angels would fear to tread. In Close's terms, the Utah researchers were "the victims of their own excessive claims" (1990, 349).

This is perhaps not surprising given the theoretical and empirical commitments of fusion physics and the clear lack of fit between those commitments and the Utah claims. In their analysis of excess heat production in the Utah electrolytic cells (based on 831 hours of on-site observation), Zeigler and his colleagues concluded, for example, that "no known fusion process contributed significantly to that excess" (Zeigler et al. 1990, 404). Even those advancing highly speculative possible explanations for the Utah findings were constrained, at least metaphorically, by traditional assumptions. Dickinson, Jensen, Langford, Ryan, and Garcia, reflecting on suggestions that cracks in the embrittled titanium enabled fusion processes, cast their analysis in the methodological terms of physics, suggesting that "crack growth results in charge separation on the newly formed crack surfaces, which act like a miniature 'linear accelerator'" (1990, 109). Although itself a departure from tradition, this account assumes a massive acceleration of deuterium ions to kinetic energy levels sufficient to allow fusion. Cohen and Davies concluded simply that "cold fusion is a known process, but not as described in the new work" (1989a, 705).

None of this should be taken to imply that a healthy skepticism of "new" knowledge is an undesirable characteristic. Indeed, in the fusion case, it seemed perfectly reasonable. The Utah claims were a stunning departure from the conventional wisdom of fusion research, grounded primarily in physics (Dye and Maugh 1989a, 28: E13). The ERAB's final report indicated that any process of cold fusion would require three fundamental revisions of traditional assumptions: a significant enhancement of quantum tunneling, drastic modifications of branching ratios in the deuterium plus deuterium reaction, and a "hitherto undiscovered nuclear process" (1989, 29–30). It thus concluded that "current understanding of the very extensive literature of experimental and theoretical results for hydrogen in solids gives no support for the occurrence of cold fusion in solids" (3).

Nearly all of the testimony before the House Committee on Science, Space, and Technology (following the University of Utah contingent) was grounded in this "current understanding," insisting that the "jury was still out" on the controversial claims of high-level cold fusion. The testimony of Brigham Young physicist Jones is representative. He characterized the search for cold fusion as "the pot of gold" at the end of a long scientific rainbow. His presentation very clearly outlined the considerable financial and technical obstacles which blocked any quick or convincing demonstration of the facticity of cold fusion. Referring to a cartoon scientist he used in his audiovisual presentation (and perhaps indirectly to Pons or Fleischmann), Jones argued that "he's trying to get a shortcut to fusion energy, but he's not going to make it. The point is, even after we can achieve yields that are comparable or greater than the energy input . . . we still have other obstacles, too; in our path to realizing fusion" (*Recent Developments* 1989, 106).

Speaking metaphorically, Jones characterized the Utah claims as a tender young plant. He concluded, "It is difficult to say what it will become. Some think and suggest strongly that this is a tree, and it will grow up very quickly and provide us enough wood for all our energy needs for generations. I do not think it is. Let's give it a chance to grow. I think adding too much fertilizer at this stage will be detrimental" (*Recent Developments* 1989, 110; see also 154–66, 189–91). In the hearings, it was clear that, aside from the relative paucity of confirmatory evidence, Pons and Fleischmann were rhetorically pilloried for lacking a properly skeptical outlook on the potential truth of their claims.

The response to Pons and Fleischmann insisted on the primacy of alternative explanatory frameworks. Broadly speaking, the topos of parsimony (Prelli 1989b) was deployed against the Utah claims so as to demonstrate their violations of the norm of skepticism. Specifically, it was argued that Pons and Fleischmann were wrong to invoke *nuclear* explanations for their observed phenomena before conclusively ruling out more conventionally accepted explanations. Such arguments abounded in the technical literature. Kreysa, Marx, and Plieth responded to Fleischmann, Pons, and Hawkins' (1989) explanation of the "liberation" of a sizable heat excess as a nuclear process. They argued that "this figure corresponds to a heat flow density of 9.26 W cm^3. The possible heat flow density due to catalytic recombination ranges from 0.12 up to 31.3 W cm^{-3}. . . . Therefore it is not really necessary to assume a nuclear process in order to explain this figure" (Kreysa, Marx, and Plieth 1989, 445). Positing a catalytic recombination, the theoretical and experimental parameters of which are more firmly embedded in the technical literature, is an explanatory account more consistent with the normative commitment to organized skepticism (see D. Williams et al. 1989, esp. 376; Lewis et al. 1989). Cohen and Davies argued that "it is wise to see whether the observations can be explained using standard physics before seeking novel solutions" (1989a, 705). The question of why particular theoretical traditions were construed as more relevant than others on this topic will be considered later.

Given that Pons and Fleischmann were generally criticized for either ignoring or misapplying other more plausible explanatory schemes to their empirical findings, they similarly were indicted for unjustifiably adopting hypothetical extrapolations as the means to that end. In effect, rather than remaining skeptical of their own anomalous findings, they were said to have "invented" a way to explain the inexplicable. The ERAB's final report was succinct: "Nuclear fusion at room temperature . . . would require the invention [not discovery] of an entirely new nuclear process" (1989, 3). Brigham Young's Daniel Decker condemned Pons and Fleischmann because, as he saw it, "rather than consider the possibility of some heretofore unknown chemical reaction being responsible, they prefer to suggest that a violation of various laws of physics is necessary" (*Recent Developments* 1989, 118). CalTech's Nathan Lewis concluded at the American Physical Society meeting, "At this point, there is no evidence for anything other than conventional chemistry. . . . We have no reason to invoke fusion to explain any of their results" (in L. Young 1989).

More specifically, consider Koonin and Nauenberg's critique of the rate of fusion claimed by Fleischmann, Pons, and Hawkins (1989). They contend that "*hypothetical* enhancements of the electron mass by factors of 5–10 are required to bring cold-fusion rates into the range of values claimed experimentally. However, we know of *no plausible mechanism* for achieving such enhancements" (Koonin and Nauenberg 1989, 691; emphasis added). The general indictment here appears to be that the Utah findings attempted to claim factual status not by secure calculations but rather by a more or less sophisticated form of wishful thinking. It is precisely that sort of thinking that should presumably be most subject to sustained and organized skepticism. Where cold fusion supporters saw "heretofore unknown nuclear processes," detractors such as Huizenga saw "miracles" masquerading as science (1992, 135). As Koonin remarked at the meeting of the American Physical Society, "It's all very well to theorize about how cold fusion in a palladium cathode might take place. . . . One could also theorize about how pigs would behave if they had wings. But pigs don't have wings" (in Browne 1989b, 22).

While the specific indictments of the Utah claims were grounded in subtly different technical concerns, what is important here is that they represent convincing evidence of the contextual deployment of the norm of skepticism. It is quite clear that there are perfectly acceptable reasons for accepting certain explanatory schemes as "better" than others. It should be recognized, however, that "parsimonious" or "empirically warranted" are relative assessments; they are not intrinsic qualities of sets of propositions. Accordingly, the processes by which they come to be taken as most accurately or efficiently "mapping onto reality" are themselves the products of communal negotiations between guiding assumptions and local instrumental goals. As such, the practical import of "organized skepticism" (and the other "constitutive" norms) is explicable only in the context of particular sets of scientific *practices*.

The evaluative implications of the rhetorical enactment of "organized skepticism" might better be illustrated via two sets of contrasts. The first involves the markedly different reaction to the Brigham Young claims, and the second concerns the rhetorical enactment of organized skepticism in the discourses of the Utah team and its supporters. Considering these contrasts also reveals the inescapably elastic nature of Merton's vocabulary and, hence, the contextually rhetorical nature of the normative structure of science.

The claims of Jones and his coworkers were theoretically as unexpected as those advanced by the Utah researchers. The enmity directed at Pons and Fleischmann, however was markedly absent in the professional responses to the results Jones and his colleagues published in *Nature* (Jones et al. 1989).[6] The *New York Times*,for example, reported that the annual meeting of the American Physical Society "physicists who have investigated Dr. Jones's report have been fairly restrained in their criticism, acknowledging that Dr. Jones is a careful scientist" (Browne 1989a, 13).[7] In this context, *careful* included the assumption that Jones's

6. Much the same can be said of the response to many fusion detractors. Huizenga argues, for example, that "these groups [reporting no fusion results] were cautious, wanting to be sure that they themselves had not made mistakes" (1992, 44).

7. McAllister (1992) suggests that Jones might also have been spared some of the vitriol because he, as a physicist, was at least a member of the community traditionally con-

claims of limited fusion effects were grounded in traditional theory, hence warranting less skepticism. As his BYU colleague, Decker, testified, "The experiment at BYU and the experiment at the University of Utah are quite different. . . . We are actually looking for a nuclear process that is already known, which people understand is part of fusion" (*Recent Developments* 1989, 114). Huizenga insists that, for good reasons, "the Jones experiments were treated from the start by the scientific community in a very different manner from the University of Utah claims" (1992, 110).

Whatever his program's intellectual grounding, it is clear that Jones rather self-consciously tempered his personal enthusiasm for his findings with a high level of (at least public) skepticism. In terms reminiscent of Watson and Crick's classic understatement in their *Nature* paper announcing the discovery of DNA's structure, Jones was quick to label his findings as simply "of scientific interest" (Jones et al. 1989). Similarly, in his testimony before Congress, he emphasized the tentative status of current fusion research: "The gap between the bona fide fusion yield and energy production by fusion is roughly equivalent to that which separates the dollar bill from the Federal national debt" (*Recent Developments* 1989, 108).

Pons and Fleischmann's claims of cold fusion, of course, were harshly criticized on the grounds that the researchers had baldly (even self-consciously) flouted the commitment to organized skepticism. There is considerable evidence for such criticism. Pons admitted unabashedly that "we believe what we are seeing. So I'm sure" (*Recent Developments* 1989, 29). Similarly, Fleischmann confirmed that "I'm still totally convinced about our own work" (*Recent Developments* 1989, 29). Such expressions of certainty were taken to suggest that all potentially competing explanations had, at least in the interim, been ruled out.

The elasticity of ostensibly constitutive commitments such as organized skepticism is perhaps more clearly exemplified in the discursive practices of the Utah scientists and their supporters. At the broad level, the norm's authority seems generally uncontroversial. While there is no equivalent account in their congressional testimony, Fleischmann, Pons, Anderson, Li, and Hawkins attribute the perception of the initial paper as insufficiently skeptical to a simple typographical error. They maintain that "the preliminary note was to have been published under the

cerned with fusion research. There is much to recommend an analysis of disciplinary affiliation in this controversy (Taylor 1991a), and I will expand this analysis shortly. Here, however, my concern is with the rhetorical accounts advanced for particular aspects of Jones's professional behavior.

title 'Electrochemically Induced Fusion of Deuterium?' but the all important question mark was omitted" (1990, 320).[8]

Despite this programmatic claim of appropriate skepticism, as a practical matter this skepticism was enacted in quite different ways in the discourses of cold fusion supporters. Most notable was the recharacterization of the findings from anomalous to novel, hence framing their heuristic value as sufficient to warrant their favorable consideration. In their lengthy, albeit inconclusive, presentation of their calorimetric evidence, Fleischmann and his colleagues observed that "the most surprising feature of these results (apart from the fact that nuclear processes *can* be induced at all in this way!) is that the enthalpy release is not due to either of the well established fusion reactions" (Fleischmann et al. 1990, 294). Similarly, Rabinowitz and Worledge of the Electric Power Research Institute, an organization with financial interests in fusion technology, argued that the cold fusion claims bore a close resemblance to less vilified claims of low-energy, cluster-impact fusion. Saying that both approaches claimed "anomalously high fusion rates based on conventional wisdom," they concluded that "both phenomena evidently need a novel theoretical approach for their understanding" (1989, 344). Note that the violation of traditional assumptions is not presumed to militate against acceptance of claims, but rather to call for additional research to explain an observation which itself is taken for granted as accurate. Venkateswaran and his colleagues insisted that the novel nuclear process posited in the Utah experiments would, of necessity, invalidate traditional calorimetric predictions. They noted that "the observed neutron bursts revealed that the fusions do not occur at a steady state. . . . hence, it is no wonder that the excess heat output went undetected" (Iyengar et al. 1990, 66).

This inverted skepticism also constrained claims sympathetic to the Utah project's neutron emission measurements. Illustrating graphically the tension between skepticism and novelty in his letter to *Nature,* Paolo challenged the wisdom of assuming that "the very small flux of neutrons generated during the experiment of Fleischmann and Pons . . . [should be] taken as proof that their conclusion is not valid." Invoking fusion processes at low levels of kinetic energy, he insisted that "if the experiments described really brought the deuterium nuclei close enough together to interact, one should expect no neutron emission and a reaction rate much higher than . . . [in] the high energy model"

8. It is important to note that Pons and Fleischmann had maintained that their observations could only be explained by nuclear processes, although "the nature of these processes is an open question at this stage" (Fleischmann et al. 1990, 321).

(1989, 711). Walling and Simons, colleagues of Pons's in the Utah chemistry department, tentatively advanced a similar argument, speculating on an "internal conversion-like process" that might produce "radiationless relaxation" of helium. Under certain conditions, they maintained, "this can explain why the bulk of energy is released as heat rather than via neutron and tritium production as one would expect in high-energy experiments" (1989, 4693).

Beyond these technical enactments of counterskepticism, the discourse of fusion supporters called attention to its practical, indeed even political, implications. On the broadest level, the traditional scientific community was characterized as being unable to transcend partisan disciplinary knowledge in order fully to consider what fusion supporters took to be the evidence. Fleischmann warned that "it's always dangerous to point at incorrect experimental data being based on theory. I think that theory should be used to explain experimental data . . . rather than just saying your data must be wrong because the theory doesn't predict that" (*Recent Developments* 1989, 28). In his preface to the Bhaba Centre studies, Iyengar identified the methodological presumptions of fusion physics as a *barrier* to knowledge production, saying that "the familiarity of scientists with accelerator-based nuclear reactions . . . led them to believe that fusion reactions can take place only on the basis of overcoming the potential barrier caused by electrostatic interaction . . . [demanding that] the particles have considerable velocity" (1990, 32).

Born of a slavish devotion to organized skepticism, this purported myopia was characterized as detrimental to the proper advance of science. In his congressional testimony, Fleischmann proclaimed that "research has been guided by the conventional approach to nuclear fusion, but it is quite clear that we would not need to be bound by that." Quite the contrary, he insisted: "We want to extend the science base . . . [and] look for the appropriate theoretical description" (*Recent Developments* 1989, 18). Representative Dana Rohrabacher (R-Calif.) commented that "there are a lot of people in the scientific community . . . who may not be open-minded towards the type of change you're suggesting is possible" (*Recent Developments* 1989, 28). Mallove, a former science writer at MIT and an avowed cold fusion believer, bemoaned what he saw as the monolithic neglect of evidence supportive of cold fusion. Decrying the "skepticism run amok that has pervaded the entire cold fusion episode," he argued that "instead of the search for scientific truth, at every turn we find people trying to shoot cold fusion down with suspicion and challenges about motives" (1991, 242).[9]

9. Mallove was responding specifically to charges aired in *Science* that a graduate stu-

As told by fusion supporters, this bastardization of appropriate skepticism was made manifest in the oppressive behavior of a threatened hot fusion community.[10] Texas A & M's Bockris and Hodko, for example, condemned the "violent denunciation of those who dare to experiment in this area. What an extraordinary travesty of the scientific approach, for *Nature* to encourage students to rubbish and ridicule the subject, not investigate it" (1990, 690).[11] Similarly, Fleischmann and his colleagues added a footnote to their technical paper in which they accused *Nature* of unjustly labeling them deceptive and then failing to rectify the impression. They commented that "much of this information was available at the time of the publication. . . . Although this was pointed out to the editor of *Nature*, he refused to publish a letter from us asking *Nature* to retract the accusations made in his editorial" (Fleischmann et al. 1990, 313).

Such vitriolic commentary (and presumably that which inspired it) was enacted in the controversy, not so much as personal nastiness, but as the result of a "skepticism run amok," as Mallove put it. In the efforts to denigrate the plausibility of the unprecedented claims and their unprecedented challenge to assumptions of traditional fusion research, it was suggested that the fusion research community had transgressed its own normative code. Worledge labeled such discursive practices of "polarization, based on too little evidence," as unscientific (in Broad 1991). In this way, the normative rhetoric of organized skepticism served implicitly (and largely by negation) to demarcate the boundaries of the domain of science.

The fact that I take healthy skepticism, in principle, as a desirable intellectual quality does not diminish the rhetorical significance of its de-

dent at Texas A&M had "spiked" electrolytic cells with tritium to ensure the experiment's success and, hence, completion of his doctorate (Taubes 1993). It is accurate to suggest that personal accusations were common in the controversy. Close (1990), for example, suggested that Pons and Fleischmann were guilty of fraud in their discussions of the radiographic evidence for fusion reactions. On this, see Petrasso et al. 1989b, 1989a; and Fleischmann, Pons, and Hoffman 1989. Mallove is not reticent about challenging personal motives himself. In a 1992 address at the University of Utah, he characterized the fusion saga as "about media myopia and media maggotry. . . . it's about honesty and falsehood; it's about misconduct, fraud, and cover-up—not by Utah, not by Pons and Fleischmann, dear friends, but by powerful, arrogant people at MIT and elsewhere" (4).

10. While certainly less common, such a charge was made occasionally by cold fusion's critics. Close, for example, wrote, "When Galileo found results that displeased the Pope, his reward was imprisonment. A group of scientists whose data did not agree with the claimed 'cold fusion' received threats of legal action" (1990, 1).

11. *Nature* and its editor, John Maddox, had been sharply critical of the Utah studies and, more specifically, of the Utah team's prepublication advocacy. See, for example, Maddox 1989a, 1989b, 1989c.

ployment in this controversy. In this case, its deployment functioned rhetorically to demarcate the causes of Pons and Fleischmann's errors from the domain of the properly scientific. Those errors were represented as the by-products of human intellectual fallibility (gullibility?) and unjustifiable speculation, and hence as unthreatening to the constitutive view of rationality around which scientific practice is organized. The version of this tale of skepticism, that told by cold fusion supporters, characterized the traditional fusion research community as intellectually subservient not to the data but to the hidebound requirements of skepticism for its own sake. That subservience, they argued, prevented a properly critical evaluation of the Utah claims and ultimately degraded the process of science itself.

It has long been assumed, publicly at least, that science is the organized attempt at "identifying, modifying, or solving problems that bear on a specific scientific community's maintenance and expansion of their comprehension of natural order" (Prelli 1989b, 122). The unparalleled amount of press coverage that surrounded the Utah fusion claims indicated that a great deal of money and fame awaited the ultimate victor in the "fusion sweepstakes." If we take Prelli's characterizations of scientific "reasonableness" and goals of science as roughly accurate (albeit incomplete), there is the clear suggestion that *personal* motivations are defined, de facto, as lying outside the realm of the truly scientific. This implicates what Merton (1973) called the norm of "disinterestedness," entailing that social actors performing the role of "scientist" abandon self-interest in favor of contributing to the development of the conceptual schemes which are taken as important for science (Barber 1952). That actor fulfills the requirements of the role when pursuing science for science's sake.

The relevance of such a normative demarcation criterion in the cold fusion controversy is substantial. If we properly understand "disinterestedness" as a rhetorical demarcation topos, then its deployment in the course of scientific controversy serves effectively to define the behaviors of particular community members as lying outside the realm of the "truly scientific." In effect, then, linking what are taken to be profound empirical errors to the "contamination" of ancillary concerns serves simultaneously to provide a rational account for those errors, which salvages the working rationality organizing scientific practice. Science, then, is rhetorically (and rather artificially) demarcated from other sets of social practice by its constructed insularity from "social" interests. Such accounting practices are clearly in evidence in the rhetorical response to Pons and Fleischmann. Indeed, Close suggests that the cold fusion saga is best understood in these terms. He maintains that "it was easy for protagonists to claim that opinions on test tube fusion, particularly its validity or otherwise, weren't always made on purely scientific grounds, but that self interests were the driving force" (1990, 51).

If confirmed experimentally, the Utah experiments would likely bring Pons and Fleischmann and the University of Utah a great deal of prestige and financial rewards. Previously, the massive capital expenditures required for fusion research had made private-sector investment in fusion projects risky and unattractive (Broad 1991). Now, however, the prospect of room-temperature fusion with decidedly simpler equipment offered hints of a viable and lucrative investment climate. Indeed, researchers from companies ranging from Westinghouse to a tiny fusion specialist, KMS Industries, rushed to capitalize on the original Utah announcement (Pollack 1989).

The Utah panel's presentation before the House Committee on Science, Space, and Technology seemed primarily oriented toward gaining government funding for their own research program and facilities to house it. Fleischmann, responding to a question from Representative Marilyn Lloyd (D-Tenn.), contended that "I would not like the Members of the Committee to think that just because we have made an initial stab at this for about $100,000 that the ongoing research will always be in units of $100,000. A high pressure steam generator, we might guess just that bit of the program . . . will cost one to ten million dollars" (*Recent Developments* 1989, 22). Later, he would conclude that "we are talking about units of tens of millions of dollars" (*Recent Developments* 1989, 34).

This call for financial support was the dominant theme in subsequent testimony of University of Utah president Chase Peterson and his hired consultant, Ira Magaziner of Telesis, USA. Peterson, an avowed disciple of "academic capitalism" (W. Charland 1989), was frank in his appeal for funding to establish a fusion research center near the Utah campus. Noting that the state of Utah had committed 5 million dollars and that the university had raised some 1.1 million dollars in private donations, he urged the panel to commit considerable funds to the Utah initiative. He argued that "the figure that comes to mind is $25 million from the Federal Government. Maybe that needs to be $125 million some day, but that's of not any importance right now. Twenty-five million dollars would allow us to start the 'onion' growing. . . . Ultimately, I would imagine it would be a minimum of $100 . . . million" (*Recent Developments* 1989, 80). Magaziner, a "business strategy consultant" (*Recent Developments* 1989, 61), lent his expertise to the cause calling for quick action so as to capitalize on the "head start" that the Utah experiments had afforded American researchers. He noted that every delay could be costly: "If it's a week from now, we may wind up losing thousands; . . . if it's a year from now, we may end up losing a couple of million. All that's not anything to laugh about. It's a lot of money. A lot of good public servants have gotten in trouble for losing track of lesser amounts of money" (*Recent Developments* 1989, 59).

While this strategy might be a perfectly reasonable adaptation to the rhetorical situation in which these claims were advanced (a congressional hearing for such a purpose), this interest would be prominently criticized in the general scientific response. In particular, the Utah team's interest in patent rights as a means of securing government support and private-sector investment was an important focus.

Not surprisingly, Brigham Young University's harsh assessment of the University of Utah's behavior most explicitly used this financial interest as a rhetorical demarcation criterion. BYU spokesman Paul Richards insisted, "We don't care about the economic ramifications. We're not interested in any patents—which only apply to any economic use of cold nuclear fusion. We care only about pure science and research" (Bernick and Jacobsen-Wells 1989). Quite clearly, there was a situationally constructed and rhetorically deployed demarcation between the domain of the scientific and that of more mundane concerns with economic gain. Close regretted that "the pressures were money and patents, and how competition of the marketplace can interfere with the more detached quest for truth in the laboratory" (1990, 13).

I do not mean to imply here that such a demarcation *does*, in fact, distinguish scientific practice from other forms of social activity. Indeed, W. Charland maintained that the boundaries between the economic and scientific have grown progressively thinner in the post-World War II era. He noted that "the presence of patent attorneys in higher education is growing at an astounding rate. . . . membership in the American Society of University Patent Administrators increased seven-fold during a recent 15-year period, from 70 to 500" (1989). Pickering's (1988) cogent account of the growth of postwar physics research demonstrates that it is impossible to conceive of the conduct of scientific research apart from its sources of funding. My point here is simply that the deployment of such a criterion to delegitimate the Utah claims (and claimants) was very much a contextual move that functioned to advance a particular (in this case, naively insular) communal identity. It is quixotic to suggest that scientists, in the course of their professional activities, can or should completely "factor out" their personal interests, economic or otherwise. Indeed, University of Minnesota physicist John Broadhurst—himself part of a team trying to replicate the cold fusion work—implicitly acknowledged the role of precisely such interests in physicists' replication efforts: "Nobody told everybody to jump on the fusion bandwagon. . . . They thought they might get some glory out of it" (in Dawson 1989, 52: F1).

This method of rhetorical "accounting for error" (Gilbert and Mulkay 1982) effectively served to attribute the erroneous claims of cold fusion to violations of the constitutive ethos of science. As a consequence, "science" was rehabilitated both as form of rationality and as social practice. Only by violating the normative order of the community, this argument suggests, could such errors have been produced and maintained in the face of mounting criticism. Huizenga argued that "the precedence of patents and copyright over scientific verification led the University of Utah to isolate itself from the mainstream of science" (1992, 55). I reiterate that this issue is analytically and practically separate from the ontological status of cold fusion. My concern is to elucidate the strategies which served to delegitimate claims about that ontological status and which constructed the presumptions of the community that would ultimately make that ontological judgment.

This should not be taken to imply either that scientists are deliberately dishonest or that they are unwitting dupes of unrecognized human motivations.[12] This analysis suggests, rather, that scientific responses to particular controversies often advance normative demarcation criteria which bear little logical relation to ongoing scientific practice. This rather self-conscious exclusion of individual interests served to distance artificially what was taken as science from other sets of social practices. Such an exclusion establishes the epistemic privilege that is routinely denied to less ostensibly pristine ways of knowing (Habermas 1972).

In the cold fusion controversy, the normative commitment to intellectual communism (Merton 1973, 273–75) is linked inextricably to that of disinterestedness, in that violations of the latter were characterized as resulting in violations of the former. Because Utah researchers maintained individual interests, they did not conduct themselves in an appropriately communal manner. Much criticism was directed at what was perceived by many as the Utah researchers' violation of this normative commitment to communal behavior. Two specific manifesta-

12. While it was ancillary to the dominant economic characterization of contaminating interests, it should be noted that intellectual obsession was occasionally mentioned as a cause of the Utah error. David Williams, an electrochemist at the Harwell Laboratory in the United Kingdom who had tried unsuccessfully to replicate the Utah radiation measurements, commented on the self-deluded state of the cold fusion "believers" ("[Con]fusion in a Bottle" 1991).

tions of this rhetorical practice were the indictment of their insufficient disclosure of relevant technical details and the indictment of their failure to abide by the structural "guarantor" of communality, peer review procedures.

Several critics charged that Pons and Fleischmann had failed (or chosen not) to provide adequate details in their initial publication in the *Journal of Electroanalytical Chemistry and Interfacial Electrochemistry* (Fleischmann and Pons 1989), as well as in subsequent appearances before Congress and several professional conferences. Oakland University physicist Craig Sevilla insisted that there were only two possibilities: "Either it's a bunch of hokum . . . or they are holding out a critical piece of information" (in McCann 1989). His colleague Craig Taylor offered a third possibility while still condemning the dearth of experimental detail, saying that "instead of deliberately taking something out of the recipe, it might have been an oversight on their part because of the tremendous pressure they were under" (in McCann 1989).

It is clear that the rhetorical boundaries of science were constructed in such a way that transgressions were portrayed as nonscientific. University of Wisconsin nuclear engineer Tom Instrator argued that "that's not the way you do science. . . . If they want [other scientists] to not give them a hard time, they should give us enough to copy it. Or maybe they have given us enough, and it doesn't work" (in Allegretti 1989). Harold Furth, of the influential Princeton Plasma Physics Laboratory, testified that he was suspicious of the motives behind this secrecy: "There is something here that he's [Fleischmann's] not talking about, and I . . . look forward to some time when his patent attorneys or whoever will let him talk about it" (*Recent Developments* 1989, 189).

Perhaps the most pointed deployment of this normative criterion appeared in a *New York Times* editorial by Lawrence Livermore National Laboratory physicist George Chapline. He insisted that the behavior of Pons and Fleischmann "is deplorable. At any time in the past month, they could have submitted the electrodes in their apparatus for an independent analysis of their helium content. . . . Their failure to submit the electrodes for analysis leaves one wondering what is happening at the University of Utah" (Chapline 1989). CalTech's Nathan Lewis expressed frustration over the inaccessibility of details at the annual meeting of the American Physical Society in Baltimore. He bemoaned the fact that "Pons would never answer any of our questions" (in Browne 1989a, 13).

Ironically, the constitutive authority attributed to intellectual communism was underscored by Pons's and Fleischmann's acceptance of it as an appropriate standard of evaluation. (Of course, they rejected claims that their own behavior contravened the standard.) Responding to congressional inquiries, Fleischmann defended their actions, insisting that "it is common practice to release a preliminary publication before you write a full paper. . . . We thought that we had given sufficient data in that preliminary publication that a cool and collected look at the paper would enable other people to replicate the experiment" (*Recent Developments* 1989, 20). In a later paper, they conceded the limitations of the original publication, writing that "it was evidently difficult for readers to reconstruct the primary information from the Tables in the preliminary publication" (Fleischmann et al. 1990, 342). Nonetheless, this was characterized not as a violation of communal openness, but as simply a misjudgment regarding the missing data's relevance to other researchers: "We point out here that a practical working system would never be based on the conditions used in obtaining the data given . . . and that statements made on the percentage of excess enthalpy . . . are arbitrary since they depend on the cell design, concentration of the electrolyte, etc." (344–45). Fleischmann also recharacterized his own lack of candor in response to further questioning regarding the withdrawal of the initial submission to *Nature,* insisting that the journal "would not be the appropriate place to submit this work in the form of a full paper; they don't publish full papers" (*Recent Developments* 1989, 21).

What is interesting here is not that Fleischmann and his colleagues differed with their critics on the proper interpretation of the Utah researchers' actions. That is to be expected. What is important is that widely disparate sets of practices could simultaneously be evaluated as in accordance with and in violation of this cultural resource. This seems compelling evidence of the elasticity and open-endedness of at least this component of the rhetorical culture of science.

The significance of these scientific rhetorics does not stop at this broad communal level, however. This alleged violation of the normative order was directly linked to the epistemic status of experimental claims because the details were considered necessary for others to attempt possible replication. Full disclosure of details is thought to offer a hedge against unanticipated or unrecognized errors in experimental procedures. *Cold Fusion Research* linked this secrecy to the inability of most laboratories to confirm the Utah findings. It concluded that the "investigators simply do not say in public reports exactly how the measurements are made; only by visiting the laboratories does it become evident that tabulations of quantities with so many [significant] figures is wholly unjustified. This practice is insidious . . . and it could explain why some groups see positive heat effects only erratically" (ERAB 1989, 48).

This rhetorical commitment was construed as sufficiently central to scientific practice that criticism of its violations even appeared in the technical literature. It is relatively uncommon for such statements to appear in technical reports, insofar as the structural constraints of journal publication actively eliminate personal references in favor of the appearance of an objective reality (Bazerman 1987; Gilbert and Mulkay 1982). For example, Kreysa, Marx, and Plieth, detailing their efforts to replicate certain critical aspects of the Utah studies, indicated that "we recalculated the data of Table 2 in ref. 1 [of the initial publication] which is not really easy since not all necessary data are given explicitly" (Kreysa, Marx, and Plieth 1989, 443). Zeigler and his colleagues, reporting on their attempts to detect charged particles in a Teflon electroanalytic cell, noted that "the above experiments are difficult to reproduce in detail since the authors discussed few details of the experimental technique" (Zeigler et al. 1989, 2929). Salamon and his coworkers extended this indictment to the level of personal communications regarding their on-site observation of the Utah cells, noting, "Unfortunately, we have not received any numerical data on excess heat production . . . so we are not able to correlate the absence of nuclear signatures with the presence of anomalous heat" (Salamon et al. 1990, 404).[13]

Ironically, this general indictment of the secretive manner in which Pons and Fleischmann reported the background of their controversial experiments is given clearest expression in relation to their avoidance of traditional peer review procedures. When the Utah team "went public," they did so without the benefit of a thorough disclosure of the technical details which peer review would presumably have demanded. Brigham Young University's Richards observed that, contrary to Pons and Fleischmann, BYU physicist Jones preferred the "*scientific* approach—submittal to a noted journal and appropriate peer review" (in Bernick and Jacobsen-Wells 1989). During his testimony before the congressional committee, Huggins of Stanford declined to specify details of his studies of radiation detection because they were scheduled for presentation at that evening's meeting of the Materials Research Society (*Recent Developments* 1989, 99).

As I have noted, Pons and Fleischmann had announced their prelim-

13. Such criticism of allegedly incomplete reporting was occasionally a feature of the discourse of fusion supporters as well. Walling and Simon (1989), for example, responding to criticism of the Utah claims, reported having received a preprint (Koonin and Nauenberg 1989) in which "enhanced fusion rates are computed and attributed to 'heavy electrons' by using a method which is essentially the same as ours." However, they continued, "no mention of internal conversion or any other mechanism for dissipating . . . excess energy as heat is made" (4695n6).

inary findings at a Utah press conference before their publication in (though not before their submission to) the *Journal of Electroanalytical Chemistry*. This was construed as a blatant violation of acceptable practice. Ronald Ballinger of MIT testified that "for results such as those . . . whose potential impact on the scientific community and the world are so great, this review is absolutely essential. Unfortunately . . . this has not happened in this case. . . . And so we in the scientific community are left to attempt to reproduce or verify a potentially major scientific breakthrough while getting the experimental details from the *Wall Street Journal*" (*Recent Developments* 1989, 178–79). Perhaps more significantly, an editorial in *Nature* condemned those "who feel compelled, either for money or fame, to ballyhoo discoveries before anyone can reasonably judge their merits" ("Cold [Con]Fusion" 1989, 362).

The rhetorical authority granted peer review functioned as a regulatory discourse that sustained the community's position of epistemic privilege as a means of securing necessary resources. Altman argued that "scientists have come to expect that communicating with the public is part of their job . . . because publicity often helps to generate funds for research" (1995).

As Frank Press, president of the National Academy of Sciences, saw it, peer review practices were central to the cultural authority of science. He suggested that "if the manner in which [Fleischmann and Pons] announced their findings is taken as a role model for others, the overall public view of science will be damaged. . . . The public will be disillusioned. The support for science will be eroded" (in Detjen 1989b, 40: D10). Arnold Rellman, then editor of the New England Journal of Medicine, concluded that "the cold fusion story is a good case in point, even if it turns out to be true. The public doesn't know what to think. There's a lot of cynicism about science, a lot of time being wasted" (in Strauss 1989b). Sociologist Nelkin expressed concerned over the Utah "science by press conference," because if the Utah claims were disproved, it might lead to the public belief that "scientific claims cannot be believed" (in Detjen 1989a, 41: A8).

The rhetorical elasticity of peer review practices, once thought field-invariant, becomes apparent when one recognizes that the revolution in electronic communications has irrevocably altered their nature and functions. Indeed, at the cutting edges of science, the currency of research consists of unpublished preprints sent via fax, Bitnet, Internet, and so on. Particularly in the sciences, research articles are often outdated upon final publication (Traweek 1988). As a consequence, the dominant rhetorical presence accorded peer review in the controversy illustrates the contextual nature of scientific practice generally and of scientific rhetorics in particular. As science policy analyst Greenberg put it, "But though often violated, like a legal speed limit, the rule of peer review remains in force and can be invoked when the establishment feels offended" (1989, 17).

The constitutive rhetorical commitment to intellectual communism expressed through the mechanism of peer review served to maintain both the rationality interest threatened by the publicizing of what most practicing scientists took to be profound error and the cultural privilege which is based on the social acceptance of that interest. To the extent that Pons and Fleischmann had violated the constitutive norm of communality, the claims they advanced no longer carried the authorizing warrant of peer review. Hence, the errors contained therein were products of ostensibly nonscientific influences. That is not to say that *real* scientists were represented as incapable of error. It is to say that those who abide by the communal peer review system are likely to have those errors identified before publication and public discussion. Profound errors made by those who self-consciously avoided peer review were said to be functions of human fallibilities and hence did not implicate the pristine rationality which is taken to undergird scientific practice. As *Nature*'s John Maddox concluded, Fleischmann and Pons "were able to nurture the delusion that they were measuring fusion only because of the secrecy they imposed on themselves" (Lemonick 1989, 75).

While the so-called public policy implications of the cold fusion controversy are not the specific focus of this study, it merits mention that the self-conscious demarcation of science and epistemic authority from broader social concerns is potentially troubling. Leaving scientific research to the "masses" is not a viable alternative. That the revelation of the human fallibility of science is seen as a threat to the authority of science, however, suggests that such authority is unreflectively based on its presumed necessity rather than its obvious social and intellectual utility. It is nothing new to observe that the "republic of science" (Boreman and Kim 1981) is not accorded universal fealty. As demonstrated in chapter 5, unreflective pronouncements of that "republic's" authority are likely to be unproductive. The chauvinistic response of Bud Scruggs, chief of staff for Utah governor Norm Bangertner, to *Nature*'s criticism of Pons and Fleischmann might serve as a gentle reminder of the implications for the artificially absolute reification of the authority of science: "We are not going to let some English magazine decide how state money is handled" (in Jacobsen-Wells 1989a, 40: E2).

This implicit totalization of science to encompass other social practices is mirrored in at least the label of the fourth pillar of the normative structure of the scientific community, universalism (Merton 1973, 270–

73). The scientific community is said to be committed to the evaluation of truth claims through discrete, impersonal cognitive criteria. As such, truly scientific knowledge claims are those which can be produced by anyone with adequate and appropriate expertise, such as a proper understanding of the appropriate cognitive criteria, methodological skills, and so on.

The constitutive implications of this norm are significant. Perhaps more than any other, it provides a normative warrant for the banishment of "threatening error" from the domain of the truly scientific. To the extent that the natural world that science is said to investigate is conceived as unitary and unchanging (though mysterious), then differences in interpretations of that reality must be a result of experimenter error. Hypothetically, at least, the commitment to universalism would entail that science does not err, but scientists do. In their moments of error, they produce nonscientific claims. This need not be taken to imply that the scientists themselves are "banished from the kingdom" of science (Pollock 1989). Scientists, after all, are human. Their claims are not. Pons and Fleischmann, qua scientists, represented no particular threat to science or scientific rationality. While they did face a number of personal attacks, such attacks were means to the ends of discrediting the genesis of their presumably erroneous claims. Those claims (and their wide dissemination) destabilized the pristine constitutive truth and rationality.

In order to understand the rhetorical constructions of universalism in the present case, it is helpful to recall that the norm entails that the knowledge products of real science are taken to be reflections of orderly natural processes (what Gross [1990a, 73] called a "privileged ontology"), untainted by the particularistic interventions of individual scientists. That errors were thought to have been made in the Utah experiments is clear. Unless the Utah claims were shown to be erroneous, the results of more than forty years of fusion research would have to be discarded, or at least modified profoundly.[14]

As played out in the cold fusion controversy, the norm of universalism was the central warrant for four classes of methodological criticisms

14. It could be argued that the Utah claims were so outlandish that they would not even call for a meaningful response. Such an argument assumes that Pons and Fleischmann were insufficiently notable scientists to warrant serious attention. I would suggest that they were at least credible enough for their claims to be taken seriously. Larry Faulkner, head of the Chemistry Department at the University of Illinois (and later a member of the ERAB's Cold Fusion Panel), attributed his willingness even to consider the issue to Fleischmann's reputation: "[Fleischmann] has won every award there is to win. . . . he's brilliant" (in Hackett 1989). Even their detractors in the cold fusion saga concede that Fleischmann and Pons had substantial scientific credentials (e.g., Close 1990, 70–74).

levied against the Utah experiments: faulty calculations, improper instrumentation, failure to ensure uniform temperature dispersion, and failure to provide adequate control experiments. These arguments functioned collectively to explain Pons and Fleischmann's errors as arising from their nonuniversal *particularized* experimental practices. Physicists found this explanation especially compelling. CalTech physicist Koonin reflected the general attitude when he said that the Utah findings were a product of "the incompetence and delusion of Pons and Fleischmann" (in Browne 1989a, 13).

Addressing Pons and Fleischmann's apparently faulty calculations, Stephen Lewis and his colleagues reported on a series of experiments performed to determine whether nuclear fusion processes could potentially occur in palladium rods which had been electrochemically charged with deuterium. They claimed unequivocally that they had not found evidence for such processes (Lewis et al. 1989, 525). They also suggested that Pons and Fleischmann's alleged evidence contained serious miscalculations: "We have also identified several subtle sources of possible error in the calorimetric measurements. If power was applied only to the load resistor until T_{cell} was attained and then the Pd/Pt [palladium/platinum] circuit was turned on and the load resistor turned off, incorrect estimates of Pd/Pt power production were obtained" (529).

Similarly, *Cold Fusion Research* took strong exception to what it characterized as "lack of estimates of precision" (ERAB 1989, 47). Noting intrinsic difficulties with the calibration and calculation of thermal gradients in fusion processes, the report identified several instances in which inadequate attention was paid to these crucial calculations. For example, the writers contended that "often, the uncertain[t]y is estimated by the precision of simply reading a voltmeter . . . when the true precision is determined by fluctuations of cell voltage or by disconnection and reconnection of terminals or other small experimental matters" (47). In this case, it was implied that the cause of the erroneous claims could be seen in the individual practices of the experimenters, abrogating their claim to universalism.

The report also reveals that the early fusion claims betrayed the commission of the most elementary calculation errors. It notes that "one sees . . . subtractions of input electrical power from evolved heat power, wherein both power quantities are tabulated to significant figures far beyond three, even when the measurements are clearly valid to no more than two or three figures" (ERAB 1989, 47–48). In this instance, it appeared that Pons and Fleischmann's particular choices had violated "impersonal cognitive criteria" of calculation, which, properly utilized, would have disconfirmed their findings.

A second general methodological argument advanced against Pons and Fleischmann involved the indefensible lack of proper instrumentation (and the requisite skill to interpret the inscriptions of the instrumentation they did have) to justify the validity of their claims. Ewing and his coworkers, for example, were especially critical of the Utah efforts to measure neutrons, a signature product of fusion. Labeling occasional bursts of neutrons from Utah-like electrolytic cells anomalous, they concluded that "these anomalies were identified as spurious detector artifacts rather than true detection because counts were not observed in the appropriate proportion in all three detectors" (Ewing et al. 1989, 404). Harwell Laboratory electrochemist David Williams, saying that he was "impressed with the rigor" of physicists' neutron measurements, conceded that most chemists "did not appreciate the difficulty" ("[Con]fusion in a Bottle" 1991).

The Utah chemists' apparent failure to appreciate this difficulty was made especially clear in the controversy surrounding their initial claims of neutron detection via gamma ray emissions, as measured by a sodium iodide scintillator. In their first article (Fleischmann and Pons 1989), a photopeak appeared at an energy range of 2.5 MeV, when established theory would predict a peak at approximately 2.2 MeV. After this discrepancy was pointed out by an MIT research team (Petrasso et al. 1989a, 1989b), subsequent publications included a peak at 2.2 MeV, but without any explanation of what Close called "the mobile peak" (1990, 279).[15]

Williams and his colleagues focused the bulk of their attention on the calorimeters utilized in the Utah experiments to measure temperature gradients. Perhaps most damning was their contention that "we built calorimeters of size and design similar to those used by Fleischmann et al. . . . We found these to be inaccurate instruments with some very subtle sources of error which it is necessary to appreciate" (Williams et al. 1989, 375). While the degree to which these researchers could in fact replicate the Utah apparatus is unclear, it is significant to note here that the locus of error was situated at the intersection of the human and the material. The "impersonal cognitive criteria" ostensibly applied by Pons and Fleischmann were said actually to carry the "taint of the human" (Pickering 1990, 685). The apparatus in question was seemingly unable to provide accurate and reliable evidence of temperature elevation because of its faulty construction by the researchers. With such practical limitations, the "empirical testimony" of the calorimeters in question

15. Some read this discrepancy as evidence of methodological incompetence, itself a damning charge. Others, including Close (1990) and Huizenga (1992), imply that Pons and Fleischmann had simply "moved" the peak, perhaps being guilty of fraud.

was rendered suspect: "Given the difficulties in operating these calorimeters, very occasional occurrences of small fluctuations cannot be considered as support for a 'fusion' hypothesis" (Williams et al. 1989, 378).

A third technical manifestation of the "universal" character of science involved criticism of the seemingly inexplicable failure of the Utah researchers to ensure a uniform dispersion of the heat allegedly produced in their "Utah tokamak." If the heavy water in the cells was not constantly stirred to mix the hot and cold water, then, even without a fusion reaction, a high temperature would be registered close to one electrode, with a lower temperature near another. Indeed, one would likely get totally different readings depending on the placement of the thermometer. Stanford physicist Walter Meyerhof insisted that such temperature differentials were most likely the result of electrolysis, the product of the electrical current breaking up the water molecules, rather than cold fusion.

Presumably, the only way to ensure that uniform heating (a signature of a fusion reaction) was taking place would be to stir the heavy water bath constantly. Pons and Fleischmann had admitted that they had not done so, reasoning that the bubbles at the electrodes would serve to "stir" the water bath adequately. Meyerhof appeared incredulous at this experimental practice. He insisted that simply shifting the location of the thermometer toward the other electrode would completely eliminate any "evidence" of fusion processes. He presented his criticism in lyrical fashion to the American Physical Society meeting in Baltimore:

> Tens of millions of dollars at stake,
> Dear Brother,
> Because some scientist put a thermometer
> at one place and not another.
> (in Browne 1989a, 13)

This appeared to be an especially damning argument, implying that Pons and Fleischmann were guilty of the most elementary experimental errors. The criticism was made even more damning by the fact that the alleged intervening factor, electrolysis, is more often discussed by chemists, like Pons and Fleischmann, than by physicists, like Meyerhof. In this sense, the locus of error was situated in particular individual failings, rather than the impersonal criteria taken to constitute science. In effect, had Pons and Fleischmann done things *like scientists,* they would not have made these errors at all.

A fourth criticism that impugned the "universal" character of the Utah claims related to the alleged failure of the experimenters to allow for or to provide adequate control conditions against which the controversial claims could be compared. Even the most rudimentary characterizations of the scientific method hold that proper experimental procedure should include control conditions (Peat 1989). Throughout the rhetorical response to Pons and Fleischmann, however, it was consistently maintained that the Utah experiment lacked precisely this condition. If, as Pons and Fleischmann claimed, fusion could be achieved with heavy water (with the deuterium isotope), then direct comparison with an electroanalytic cell with "light" water should presumably manifest different levels of heat production. Furth, of the Princeton Plasma Physics Laboratory, testified before the House Committee on Science, Space, and Technology: "If I were Sherlock Holmes, I would refer to this as 'The Case of the Missing Control Experiment' and I would ponder what it meant" (*Recent Developments* 1989, 168). What it "meant" was articulated in *Cold Fusion Research:* "In this circumstance, it is *not acceptable* to compare results from many test cells against the behavior of one, or only a few, control cells. Statistically meaningful results *require* comparisons of like numbers of cells of both types" (ERAB 1989, 47).

Significantly, the results of the Utah experiment were deemed as unacceptable, not merely suspect or inexplicable. If, as I have argued, the standard of acceptability operative in this context was that of providing a rationally consistent explication of a rationally consistent empirical world, then the absence of control conditions defined the Utah findings outside the domain of even potential scientific credibility. Accordingly, the significant threat to the accumulated wisdom of fusion research was "accounted for" by reference to the inability of the Utah claims to enter the corpus of scientifically legitimate (and warranted) knowledge.

These specific experimental criticisms also contributed to the rhetorical negotiation of the nature and definitional importance of replication in scientific practice. Recall that universalism entails, in principle, that any competent researcher should be able consistently to verify the results of an experiment by invoking impersonal criteria. That is, he or she ought to be able to replicate it. The requirement that particular experiments be capable of reproduction has long been considered a defining characteristic of science (H. Collins 1975, 1982a, 1982b, 1984, 1985; Franklin and Howson 1984; Travis 1981). To say that science, as a social practice, is actively and consistently committed to a simple conception of replication, however, both trivializes the notion of replication and misrepresents scientific practice. What we often take as an intrinsic characteristic of science, replication, is itself as much a contextually deployed rhetorical resource as the sorts of arguments noted above.

Underscoring its rhetorically contextual nature is that, in normal scientific practice, replication is rarely accomplished at all (G. Bowden

1985; Travis 1981). Priority of discovery, in many ways, is a more honored accomplishment in the scientific community (Dye and Maugh 1989a). Harry Collins points out that replication is less a consideration in everyday scientific activity than a "useful fiction" which can be advanced to serve localized technical or professional interests. He maintains that "though scientists will cite replicability as their reason for adhering to belief in discoveries, they are infrequently uncertain enough to . . . press this idea to its experimental conclusions. For the vast majority of science replicability is an axiom rather than a matter of practice" (1985, 19).

Replication functions axiomatically in the sense that it is invoked only when particular professional or technical interests are called into question. In such cases, it breaks potential conceptual or evidentiary impasses. The reliance on the topos of replication is not necessary or intrinsically scientific. It is a rhetorical act, then, addressing certain contextual exigencies (Prelli 1989b, 160–67).

To suggest that replication is bound to the circumstances in which it is made rhetorically present is equally to suggest that what counts as a replication is situationally variable as well. Since, in the spirit of Heraclitus and his river, it is temporally impossible to actually do the *same* experiment over, replicability implies that there is some independent standard by which one can locate a competently performed (hence, scientific) replication (Pickering 1981; Travis 1981). Of course, in crucial ways, that which is taken to "count" as a competently performed replication is inextricably tied to its own outcome. As Collins points out in his discussion of the replication of the gravity wave detection experiments, "We won't know if we have built a good detector until we have tried it and obtained the correct outcome! But we don't know what the correct outcome is until . . . and so on *ad infinitum*" (1985, 84). Similarly, Pickering argues that "one cannot separate assessment of whether an experimental system is sufficiently closed from assessment of the phenomena it purports to observe" (1984a, 109).

The discourse of the cold fusion controversy gives ample evidence of its struggle to manage the contextual variability of replication rhetorically. That the charge of irreproducibility was raised *against* the Utah experiments is clear. Pool, writing in *Science,* noted that "the first problem to be overcome is that of reproducibility. For whatever reasons, both the neutron emissions and the excess heat are difficult to pin down and even laboratories that have seen the effects cannot repeat them in every test sample" (1989a, 1040). Others reported numerous and sophisticated attempts to replicate the Utah experiments so as to determine whether the controversial knowledge claims were supported, or supportable (D. Williams et al. 1989; Gai et al. 1989; Kreysa, Marx, and Plieth 1989; Lewis et al. 1989). For these critics, failed attempts at replication meant that the Utah experiments were insufficiently universal and, by implication, unscientific (Chapline 1989; ERAB 1989, 41–43; Pool 1989c; *Recent Developments* 1989, 189). As Huizenga concluded, "This irreproducibility and sporadic character of cold fusion . . . has served as one of the grounds for the skeptics to reject cold fusion as valid science" (1992, 87).

The variability with which the replicability criterion was deployed in the cold fusion controversy reinforces its rhetorical character. As often as the charge of irreproducibility was raised against the Utah experiments, it was deployed with equal fervor (but in contradictory ways) by the Utah researchers and their supporters. Of course, the protagonists claimed that the original experiments had, in fact, been replicated. Pons, for example, testified that measurements had "been made many times and by many different . . . methods" (*Recent Developments* 1989, 13). The initial report of the results described four separate experiments which the researchers claimed produced consistent conclusions (Fleischmann and Pons 1989; Fleischmann et al. 1990).

Beyond these claims, however, Pons and Fleischmann maintained that the charges of irreproducibility against the Utah procedures were themselves the products of particular experimental lapses. Fleischmann testified, "I've expressed the view repeatedly that any scientific process requires independent verification. . . . I have been given access to some people's experimentation who have not been able to find the heat, and which has been totally unsurprising to me" (*Recent Developments* 1989, 22). He later insisted that the cause of these failures to find similar results was to be found in the replication procedures. He argued that "they would never have been able to find it using the apparatus they have used" (22). In an extended tu quoque, the Utah team argued that the replication attempts of their detractors were evidence not of the invalidity of fusion claims but of the experimental incompetence of their detractors.

Beyond this, though, such contradictory articulations of universalism, in the form of replication, suggest the thoroughly rhetorical nature of what have traditionally been taken to be constitutive elements of science. Traditional conceptions of replication (i.e., as a regular, relatively predictable phenomenon) were problematized by fusion boosters. Mallove summarized the issue in these terms: "To an extent, the phenomena remain not repeatable at will—but repeatable to be sure, in a statistical sense, and sometimes now with very high confidence" (1991, x). At the first annual Conference on Cold Fusion in Salt Lake City,

Preparata responded to *Nature*'s conclusion that the Utah results were unreproducible, arguing that "the fact that you cannot reproduce tritium doesn't mean that it's not there. . . . You have to do science without arrogance and with patience" (in Mallove 1991, 223). Bockris and Hodko of Texas A&M insisted on the veracity of the Utah findings while conceding that "one cannot produce the phenomenon on demand, unlike, for example, electroluminescence" (1990, 688). Howard Menlove of Los Alamos National Laboratory concluded that "we need to change our vocabulary on reproducibility. . . . Earthquakes are believable, even though they are not reproducible. Tremors go on all the time. If we sharpen our techniques . . . we'll be on our way" (in Mallove 1991, 231).

This apparent lack of reproducibility (as traditionally understood) also constrained technical attempts to account for the Utah findings. Lin and his colleagues, for example, advanced a "mechanism speculation" based on the formation of dendritic protrusions on palladium after prolonged electrolysis. They then suggested that fusion occurs on the surface states of peaks created by the buildup of nickel, iron, and chromium. They hypothesized that the "necessary preliminary event for tritium production is the substantial growth of the promontories described. Such growth would be lengthy and *irreproducible* because the necessary impurities would arise as a result of, e.g., dissolution of the anode, diffusion from the glass, etc." (Lin et al. 1990, 208; emphasis added).

Indeed, some supporters inverted the probative value of the replication standard, arguing that it was precisely the lack of reproducibility (on demand) that offered a definitive hedge against a major indictment of the Utah studies—that tritium contamination of the electrolytic cells actually accounted for the results. Bockris and Hodko maintained that "the formation of tritium observed in cold fusion experiments has . . . several characteristics which rebut contamination arguments. . . . It is not reproducible, as is the natural enrichment of tritium owing to electrolysis" (1990, 689).

This, of course, raises the question of which authorizing audience is deemed as having legitimate authority over a research area which had clearly been expanded beyond its traditional affiliation with fusion physics. In such a case, internal demarcations of the domain of the truly scientific are not likely to prove conclusive because both potential communities might rhetorically articulate similar programmatic commitments. So the preservation of those rhetorical commitments cannot alone conclusively determine the outcome of the cold fusion debate. To suggest otherwise would be to beg the question as to whose specific experimental/theoretical presumptions are to be brought to bear in the final determination of facticity.

That is a question which seemingly betrays the unnecessarily restricted (and restrictive) traditional view of the demarcation of science. There is a great deal of value in explicating the rhetorically emergent constitutive norms of "science." That study, however, cannot tell us everything of relevance about the dynamics of demarcation. There would appear to be important cases, such as cold fusion, though, in which competing research communities might generally accept the same rhetorically constitutive normative commitments but differ on the dimensions of scientific practice which might conceivably be brought to bear on competing knowledge claims. It seems possible that important interdisciplinary demarcations might also be drawn so as to maintain and/or challenge the relative warranting authority of particular research communities and, in effect, extend the ecosystem enacted in the cold fusion saga. It is that possibility that is explored in the final section of this chapter.

Extending the "Ecosystem" and Closing Down the Debate

The above analysis has clearly indicated that rhetorical demarcation practices functioned in the cold fusion controversy to construct complex accounts for the errors committed by Pons and Fleischmann in the production of their claims of high-level cold fusion. This rhetorical construction of the constitutive normative structure of science, however, does not provide a full understanding of the relationships between the social practices of science and the broader social structures in which they are embedded. To stop at this point would seem to warrant a demarcation of science which *is* conceptually and practically distinct from those broader social structures. Accordingly, I want to consider the possibility that, while certain conceptual criteria (e.g., the "norms") which are negotiated as constitutive of science might be less obviously connected to broader social influences, a brief study of interdisciplinary demarcation practices will more centrally reveal those influences. The normative conceptual parameters of science are given their contextual meaning in scientific practice. Scientific practice, in turn, is heavily constrained by constitutive factors typically thought to lie outside of science.

In the cold fusion controversy, an internal demarcation line was rhetorically constructed between the competing research communities of fusion physics and electrochemistry. This interdisciplinary demarcation can be understood as serving important professional and technical interests which were threatened by the increased public currency of the fusion claims (Taylor 1991a; McAllister 1992). More significantly, I will suggest that this was an epistemically consequential demarcation, insofar as it actively constrained the application of particular theoretical frameworks and research exemplars.

While some limited research on cold fusion continues, most notably in Japan (Pollack 1992; Freeman 1992), and definitive conclusions on its ontological status remain controversial (Broad 1991; Port 1992), it appears safe to say that a fairly firm consensus (in both research communities) appears to have developed against the controversial claims of high-level cold fusion. It could be argued that the controversy was "settled" simply because "truth won out"; that is, one side was right, and the other side was wrong. This seems a rather ingenuous explanation, however, in considering the early phase of the controversy, when nothing approaching a crossdisciplinary consensus had developed on that question. More to the point, such a position does not deny the clear demarcations which were drawn early in the controversy. Indeed, Maddox seemed to confirm precisely this distinction when he lauded the fact that "the brief spell in April when it seemed as if cold fusion would permanently divide chemists and physicists has left no trace" (1989a). Similarly, Clark argued retrospectively that "chemistry and physics are not very different sciences. The overlap between them is enormous and growing. The cold fusion reports [however] seemed to polarize these two fields like nothing in recent memory" (1991, 277). I want to speculate here on the motives of that "division" and its rhetorical manifestations.

In order to appreciate the conceptual professional interests at stake in the cold fusion controversy, it is important to recall that fusion research had traditionally been considered the professional domain of physics and physicists. Clearly, the claims announced at the Utah press conference served as a compelling challenge to this presumptive division of experimental labor. For years, physicists had sought to harness fusion power by subjecting hydrogen atoms to immense pressures and temperatures inside tokamaks or laser bombardment devices, which might easily cost up to ten million dollars per year to operate (Strauss 1989b; *Fusion Energy Program* 1989, 342–46). But now two chemists claimed to have discovered ambient fusion, using equipment that was available in most high school chemistry laboratories, in a basement laboratory while spending approximately one hundred thousand dollars over a five-year period (*Recent Developments* 1989, 11–12). (Pons admitted, however, that since university facilities and utilities had been utilized, that figure was slightly lower than actual expenditures.) Science policy analyst Greenberg observed that "offense was keenly felt by a particularly powerful sect of science, physicists, who for decades have been working on 'hot' fusion, on a current annual budget in the United States of $350 million" (1989).

This "offense" was often articulated as a manifestation of proprietary interest in the topical domain. University of Illinois researcher George Miley said rather flatly that fusion was appropriately situated in the "field" of physics. While rejecting the Utah claims, he testified that "I am personally convinced that solid-state catalyzed cold fusion occurs and this is an unexpected and very important new regime of *physics*" (*Recent Developments* 1989, 131; emphasis added). I think it rhetorically interesting to note that fusion was identified as important not for "science," generally speaking, but rather for physics in particular. I do not mean to suggest, of course, that Miley (or anyone else) is so intellectually provincial as not to recognize the more general implications of fusion research. Indeed, Miley himself is a professor of nuclear and electrical engineering, rather than a physicist. What I want to suggest is that fusion research traditionally (and presumptively) had been considered to be in the domain of physics, by physicists and others like Miley. As Princeton's Furth put it, "We feel, at least as physicists, that maybe the chemists should . . . not tell us physicists that we need to change our physics to explain the [fusion] process" (*Recent Developments* 1989, 115).

Given the development of disciplinary affiliations, limited conference travel support, and so on, it is not at all surprising that the professional interaction between the different disciplines at work on the problem was limited (Suarez and Lemoine 1986; Schmidt 1989). Indeed, after the claims of cold fusion were announced, this became painfully clear. The University of Utah vice president for research, James Brophy, explained Pons and Fleischmann's decision not to attend the American Physical Society meeting in these terms: "That's like putting David *[sic]* in the lions' den. They're not members of the American Physical Society—they are members of the Electro Chemical Society" (in Christian 1989, 52: F13). BYU's Richards indicated that physicist Jones "doesn't usually attend conventions of chemists, but . . . will this time" (in Christian 1989, 52: F13).

While one hesitates to place too much emphasis on such statements, it is clear that, particularly for some chemists, the Utah announcement represented something of a "coming of age" for research efforts in chemistry. Suffering from a perceived poor sibling relationship with other scientific disciplines, most notably physics, chemistry caught a glimpse from the Utah results of a future of public approbation (Beall and Berka 1990; McAllister 1992, 25–27). As a result, fusion supporters were not particularly reticent in proclaiming what they took to be their new-found authority. One of Pons's University of Utah colleagues, Walling, defended the Utah team's experimental procedures in the face of criticism from physicists, saying, "I guess it's not fancy enough for them" (in Schmidt 1989). During an address at the meeting of the American Chemical Society, ACS president Clayton Callis drew laughs when he opined that traditional fusion programs were "too expensive and too ambitious to lead to practical power" and that "now it appears that chemists have come to the rescue" (in Close 1990; Pool 1989d, 284). Fleischmann, testifying before Congress, appeared a bit more magnanimous. He suggested that "the people who have got the big experience in the high energy physics end will have an absolutely vital contribution to make. I think they will come to see that very shortly" (*Recent Developments* 1989, 29). Grabner and Reiter (1979) have argued that physics researchers are often conceived of (and often conceive of themselves) as the "guardians at the frontiers of science," occupying a privileged position in the scientific community. In one sense, then, the Utah claims challenged that privileged position and its attendant degree of epistemic authority.

There are compelling reasons to accept this interdisciplinary demarcation as epistemically consequential, at least in the early stages of the cold fusion controversy. While I take no position on the ultimate "truth" of the Utah claims, I suggest that the rhetorically sustained interests of the competing research communities actively constrained their early empirical evaluations of those claims. Clearly, assessments of the Utah claims did not adhere strictly to disciplinary lines (McAllister 1992). For example, two of cold fusion's sharpest critics, Nathan Lewis and David Williams, are electrochemists—Lewis at CalTech and Williams at Harwell. Numerous and complex reasons, stated and otherwise, constrained individual evaluations of the fusion phenomena. I simply suggest that the rhetorical construction of interdisciplinary lines of demarcation *also* informed and influenced those evaluations. While hardly an impartial observer, Mallove noted that "throughout the cold fusion saga . . . chemists were as a rule more disposed to believe Pons and Fleischmann than were physicists" (1991, 71).

In a very broad sense, this rhetorical construction of interdisciplinary demarcation lines actively constrained the appropriate experimental framework for the evaluation of cold fusion. Pool characterized the fusion physicists' perspective when he wrote that "we know what fusion looks like, we know what it takes to produce fusion, and this isn't it. If something is there, it must be a chemical reaction" (1989d, 284). Recall that the natural repulsion of two deuterium nuclei presumably would necessitate high levels of energy to be overcome. Consequently, Pool argued that for physicists, it was "difficult to see where this energy is coming from" (1989d, 284–85). Reflecting sardonically on the ACS meeting in Dallas, Huizenga wondered, "Might it not be more profitable for this audience of chemists to have had an additional lecture in basic nuclear physics in order for them to understand the problems associated with Fleischmann and Pons's claims?" (1992, 32). What is of particular concern here is that the assumption of physics' explanatory primacy, in many ways, would have preordained the evaluative closure of this debate before it ever began.

The chemists' perspective, on the other hand, appeared to be this: "We know what chemical reactions look like, and there is no possible chemical reaction that could be producing this much heat. It must be fusion" (Pool 1989d, 285). As Allen Bard, a University of Texas electrochemist (who would later recant and serve on the fusion advisory panel of the ERAB), argued, "The lesson that more heat is produced than can be accounted for . . . is starting to get through to me. The effects are starting to add up to a fairly strong case" (in Pool 1989d, 285). Similarly, Pons, testifying before Congress, was succinct in his response to physicists' criticism of his publication choices. It is important to note that the response was articulated along disciplinary lines: "Physicists don't do things exactly the same way [as chemists do]. . . . This is simply a different system that chemists do not use" (*Recent Developments* 1989, 20).

It seems clear that disciplinary identification initially influenced determinations of what was taken to "count" as relevant evidence for the facticity of cold fusion. I previously discussed the relative emphasis placed on the absence of control conditions in the Utah experiments. Furth of Princeton linked the evaluative importance of this factor to a disciplinary demarcation. He maintained that only the performance of control experiments would galvanize the *physics* community into belief and action (Peat 1989, 93; Pool 1989b; Maddox 1989a). Gai and his colleagues focused primary critical attention on the presence (or absence) of radiation measurements. They insisted that "nuclear-fusion events *must* be characterized by the production of the ionizing radiations . . . as there is no known mechanism that inhibits their emission" (Gai et al. 1989, 29; see Chapline, in Pool 1989d, 284).

The question of the "missing radiation" proved a troubling one for most physicists. Robert McCrory of the University of Rochester advanced what he called the "dead graduate student problem." McCrory calculated that the production of the claimed four watts of energy would require trillions of fusion reactions per second. Because each fusion reaction produces a neutron, the laboratory should have been showered with lethal levels of neutron radiation. He reasoned that if Pons and Fleischmann had been correct, then the graduate students tending their laboratory should have been killed (Peat 1989, 82–83; Bishop 1989a).

Chemists, however, from a different disciplinary perspective, rejected that criticism. Pons, for example, noted that fusion physicists were accustomed to conceiving of fusion rates in fractions of seconds at enormously high temperatures. The Utah reactions, on the other hand, had been said to have taken place over a period of two hours inside the solid palladium crystal. Pons maintained that "there's no reason the reaction [in the palladium] has to be the same" as the reactions occurring in a more traditional fusion research apparatus (in Bishop 1989a). "Providing answers to safety concerns," Miles and Miles insisted that "radiation exceeding weekly safe limits would not be reached even if the cold fusion bottle were held against the body for an entire 40-h work week" (1990, 412, 411). A more technical explanation was advanced by Walling and Simon. In terms consistent with Paolo (1989), Walling and Simon proposed a "radiationless relaxation (RR) path . . . in which energy is transferred to the PdD_x lattice, perhaps mediated through the lattice electrons." They argued that such a mechanism would be "attractive from the point of view of heat and energy production, since it predicts that each fusion event could produce up to 24 MeV of heat, unaccompanied by a large, troublesome neutron flux or ^3H formation" (1989, 4694). The so-called actual empirical value of the arguments aside, this can be understood as a practical negotiation of competing (discipline-bound) understandings of those empirical values.

This brief analysis suggests that rhetorical processes of demarcation reified taken-for-granted, field-dependent understandings of the nature of chemical and physical processes. These understandings facilitated the early conflicting evaluations of the fusion claims. The relevance of this issue lies not in revealing the dynamics of interdisciplinary skulduggery or interpersonal unpleasantness. Rather, it lies in the indication that, at least in the early stages of the controversy, the politics usually assumed to be exported from science were omnipresent, serving ironically even to structure the demarcation discourses that attempted to export it.

We can now follow the ecosystem outside the boundaries crafted explicitly in the controversy. Why, given the sharp disagreements regarding the appropriate disciplinary framework for evaluating cold fusion, did the physics community ultimately prevail? To find answers, I think, we must look not to the emergent ecologies of Salt Lake City and College Station, nor to Princeton and Palo Alto. Rather, we must turn our critical gaze toward the perturbations of the ecosystem emanating from Washington, D.C.

In addition to the conceptual and evidential constraints exerted by this disciplinary demarcation, there was also a sense in which these were reembedded in the material practices of modern science. Hence, a second crucial interest at stake in this controversy in scientific practice related to the material support for the pursuit of particular disciplinary research exemplars. Clearly, the historical linkage of federal research funding with the multimillion-dollar physical research devices were, at least in theory, threatened by reports of an inexpensive, relatively uncomplicated method for achieving cold fusion. This was especially the case in an age of fiscal belt-tightening or, at least, of the perception that such might be a good thing. In 1989 federal fusion support totaled some five hundred million dollars, with approximately 70 percent of that allocated to magnetic confinement research (Crawford 1989; *Fusion Energy Program* 1989). Representative Rohrabacher crystallized the tension between conflicting funding applications: "Perhaps the fact that so many people in the scientific community are now dependent on Government grants, that perhaps are heading in totally the opposite direction to achieve the same results, might actually make this problem even worse" (*Recent Developments* 1989, 29).

The incremental progress and bright prospects in fusion physics research might have been readily apparent to practicing scientists. Testifying several months after the fusion saga began, Furth, for example, proclaimed that "during the past decade, progress in magnetic fusion research has been very strong. . . . The operation of a fusion power demonstration facility is a realistic goal for the first quarter of the 21st century" (*Fusion Energy Program* 1989, 25). Such progress, however, was much less apparent to those structures which allocated the lion's share of funding. Pons and Fleischmann (and, significantly, President Peterson and his consultant) had not come to Capitol Hill for a courtesy call. They had requested an *initial* outlay of some $25 million, growing to $125 million, for the establishment of a cold fusion research center at the University of Utah (*Recent Developments* 1989, 80).

If such funding requests had been granted—but they were not (see ERAB 1989; and Lindley 1989b)—the nature of externally funded scientific research in the United States would have been changed dramatically. Indeed, there was a recognition that virtually no Department of Energy funding had gone to alternative approaches to fusion research (Chandler 1989b, 65:G2). While one would not want to suggest that the tokamak would immediately become the "white elephant" of the scientific community, it is apparent that a research tradition organized by, and lucratively funded for, the physics community would be altered fundamentally. That this alteration would be based primarily on an interdisciplinary demarcation also seems clear. Just as the attribution of professional authority over fusion claims to the physics community in the post-World War II United States had been facilitated by unparalleled funding from government sources, so a newly organized fusion research program would claim a large measure of that authority for electrochemistry and those fields with which it was "allied" (Latour 1987). It is at this point that interdisciplinary demarcations most directly influence the actual conduct of science, or shape scientific practice.

Such a "rhetorical situation" necessitated a rhetorical response which would either reshape or reaffirm the sociopolitical "system" which had made concrete the previous demarcations of interdisciplinary authority. Cold fusion supporters, not surprisingly, were sharply critical of the current funding and peer review structures because, as they saw it, they did not promote creative research (i.e., certain of their own projects, including cold fusion). Pons straightforwardly repudiated the more "restrictive" funding sources which currently organized scientific practice in favor of those offering more freedom as well as a less stringent implementation of the epistemic hierarchy. He turned down a $322,000 Department of Energy grant in favor of continuing a grant from the Office of Naval Research which had not been specifically targeted for fusion research (Fitzpatrick 1989). Office of Naval Research spokesman Bill Lescure indicated that scientists are given considerable leeway in what they choose to investigate under ONR guidelines (in Fitzpatrick 1989). Such claims seemed to have fallen on at least a few sympathetic ears during the April hearings. Representative Robert Packard (R-Calif.) labeled the current peer review system of allocating federal research funds as "incestuous," calling it a "type of arrangement where the money goes to those who make the decisions" (*Recent Developments* 1989, 83). He subsequently bemoaned, with support from Utah president Peterson, the fact that some researchers "almost by nature of the structure, were locked out of any opportunity to begin, that they couldn't get into the entry level of some of the funded programs" (84).

In sharp contrast, traditional fusion researchers graphically underscored their commitment to the current funding structure. Calling cold fusion a "wildcat oil well" (*Recent Developments* 1989, 129), Miley insisted that "I do not believe that cold fusion should force any near-term changes in funding for the older established fusion research programs (i.e., magnetic or inertial confinement fusion). We simply do not know enough about cold fusion yet" (*Recent Developments* 1989, 146). It is interesting to note that Miley rather explicitly grounds his call for maintaining the structural status quo in an interdisciplinary demarcation. He seems to suggest that funding the *practice* of science should necessarily follow from a particular epistemic hierarchy in which warranting authority is primarily vested in the precepts of physics. Similarly, Furth warned that "continued reprogramming of funds from the magnetic-confinement effort into other areas is likely to have a significantly damaging net effect on the strength of the U.S. fusion effort" (*Recent Developments* 1989, 175).[16]

Ultimately, it appears that this demarcation-based hierarchy was affirmed. The Energy Research Advisory Board, which reported to the secretary of energy, unequivocally rejected substantive alteration of current scientific practice. The final report advised "against the establishment of special programs or research centers to develop cold fusion. . . . The Panel is . . . sympathetic toward modest support for carefully focused and cooperative experiments within the present funding system" (ERAB 1989, 1).

This broadly sketched relationship between interdisciplinary demarcation, funding practices, and the conduct of scientific research suggests that many of our traditional assumptions about the separation of the political (even ideological) and the scientific may be in need of revision, if not outright rejection. I argued earlier that static, transcendent demarcations of "science" are generally useless for characterizing the conceptual boundaries which are invoked to demarcate it from other social practices. Here I would argue that conceptualizing science as a social practice forces the sensitive critic to recognize the intrinsic and constitutive influences of political and economic factors on what it means to "do science."

At this point, then, it should be apparent that these disciplinary rhetorics were both constrained by and exerted transformative pressure on a wide range of communal practices. However we ultimately evaluate the cogency of the Utah claims (and there seems to be considerable reason to doubt that cogency), what is still unclear is how, precisely, these disciplinary rhetorics are connected to the broader political practices to which I alluded earlier.

The connection, I think, is born of the capital-intensive nature of contemporary techno-science. While the traditional image of the scientist puttering away in her or his laboratory, untouched by the vagaries of

16. Before the 26 April hearings, the committee's Energy Research and Development subcommittee had recommended, on request of Robert Walker (R, Pa.), that five million dollars be reallocated from the traditional fusion research budget to support cold fusion research. The decision was reversed later by the full committee.

budgets, provides grist for Hollywood and for science textbooks, it bears little connection to contemporary techno-science. Latour aptly described the material ground of scientific practice, saying, "When scientists appear to be fully independent, surrounded only by colleagues . . . it means they are fully dependent. . . . when they are really independent, they do not get the resources with which to equip a laboratory [or] to earn a living" (1987, 158). The apparatuses which form the basis for so much of our advanced scientific research are made possible only through a relatively stable supply of research funds, whether from corporate interests or from the federal government. Put simply, science and scientists cannot "just do it." To an increasing extent, they resemble Tennessee Williams' faded grande dame, Blanche DuBois, dependent "on the kindness of strangers."

While kindness might have been less a consideration than technical viability, the requisite funds for cold fusion research were not forthcoming. In this case, then, the ultimate decisions to deny funding for concerted cold fusion research programs effectively doomed their future development, whatever the ontological status of cold fusion. Without the requisite materials with which to expand the experimental base and, in the (however unlikely) event of positive results, with which to produce energy efficiency, the cold fusion program could not possibly succeed. As Peat put it, "The bottom line on fusion power is a matter of economics. Fusion laboratories—whether they be tokamaks, magnetic mirrors, or lasers firing at frozen pellets—are megaprojects and require megadollars" (1989, 36).

While the Utah team's request for twenty-five million dollars pales by comparison with budgets for projects like the supercollider, those so-called megadollars are at the heart of laboratory practices. Huggins lamented that he was forced to pursue his research, "with as much as vigor as we can, recognizing the limitations of zero funding" (*Recent Developments* 1989, 102). Referring to the future prospects for cold fusion-based technology, Fleischmann claimed that "any development of scale up is totally limited by the cash flow" (*Recent Developments* 1989, 25). Decker, calling for maintenance of the current funding structure, offered a direct illustration of the interaction of the rhetorical and material dimensions of the controversy. He maintained that "in order to see the fusion, very sensitive neutron detectors are required which are not available in most laboratories. These are not expensive, but they are rare. . . . I would think ten to twenty million dollars . . . would be adequate" (*Recent Developments* 1989, 126). That the detectors are not expensive (by the standards of big science) is of little consequence. Whatever the cost, the key issues are that they are prerequisites to knowledge production and that their acquisition largely depends on rhetorical transactions occurring outside the laboratory.

Accordingly, the ostensibly political deliberations which yielded the judgment to deny funding should not be written off as nonscientific, or even as only marginally scientific. It seems to me, rather, that the conduct of laboratory practice was directly and inherently tied to their outcomes. This is not to say that Robert Roe (D-N.J.), chair of the Committee on Science, Space, and Technology, donned a lab coat and measured neutron flux. It is to say that laboratory practice is but one constituent of the cultural (and rhetorical) practices of science in the modern age. As the *Congressional Quarterly Weekly Report* concluded, "Cold fusion . . . is a . . . political phenomenon . . . subject to the stratagems, tradeoffs and turf wars that accompany politics—and Congress is right in the thick of it" (Cloud 1989).

This is not to say that the cold fusion saga was in fact a case study in how nonscientific behavior sullied the public ethos of real science. On the contrary, I think it serves to alert us to the inescapably human dimensions of real science, so that we might appreciate its strengths without wishing away its imperfections. In Pinch's terms, "In cold fusion, we find science as normal. It is our image of science that needs changing, not the way science is conducted" (1994, 99). Understanding demarcation, the practical construction of what it means to conduct science in the first place, seems especially suited to enacting that change.

POSTSCRIPT

STABILIZATIONS AND PROVOCATIONS

From a sociological perspective, John Ziman (1978) characterized the attempt to understand the nature of science as equally presumptuous as attempts to understand the meaning of life itself. From a philosophical perspective, Larry Laudan (1983) labeled the demarcation of science a "pseudo-problem." From a rhetorical perspective, this study has portrayed the demarcation of science as neither presumptuous nor pseudo. Viewed in my terms, demarcation is a practical, rhetorical accomplishment with important social, epistemic, and political consequences. Those consequences highlight the theoretical and practical significance of understanding science as a complex "ecosystem" of interpenetrating constituents and practices. In this postscript, I want to summarize the theoretical and critical arguments developed throughout this study, indicate their theoretical limitations, and discuss briefly the theoretical and critical implications.

So far, I have developed a view of demarcation as a practical matter which is rhetorically negotiated by scientists in ongoing scientific activity. I have not argued that scientists speculate consciously on the philosophical and political foundations of their daily practices. Rather, a rhetorical perspective portrays demarcation as a discursive accomplishment of those practices. This perspective develops and is premised on several assumptions about the nature of science. First, science is one among many sets of social practices. Second, as with all meaningful enterprises, scientific practice is organized by particular interests: personal, cognitive, technical, professional, and so on. Third, scientists' understanding of the boundaries of their social world is socially constructed in and through discourse, particularly in situations in which those interests are challenged. Scientists, consciously or otherwise, rhetorically construct operative definitions of science which serve to exclude what they take to be nonsciences or pseudosciences, in order to enhance their relative cognitive authority and to maintain a variety of professional resources, such as limited funding or control of school curricula. Beyond these distinctions between science and nonscience, this study has also argued that demarcation is accomplished when competing research communities within traditional science construct working definitions of *appropriate* science in order to advance proprietary interests over particular research domains and/or control of limited material resources.

The development of this theoretical position began with a rhetorical reconstruction of previous approaches to demarcation issues in chapters 2–4. Acknowledging that distinctly different questions about demarcation have been posed in philosophy and sociology, I appropriated these discussions for heuristics helpful in constructing a rhetorical approach to the demarcation of science. It was first necessary to abandon the notion of science as a discrete set of propositions which bear certain relationships to each other or the method by which they are validated, in favor of a view of science as practice. Second, there was a recognition of the intrinsically social nature of that practice. What scientists do, how they do it, and how they evaluate what they have done are all bound to processes of social and communal authorization. Finally, this focus on the social practice of science served as an entry to a view of science as rhetorical practice, in which the social meaning of science is variably constructed in different rhetorical situations and is enacted in alternative configurations of persons and practices.

Case studies of recent disputes about creationism and cold fusion illustrated the implications of the rhetorical choices constraining the locally stabilized ecosystems. As consequential discourse, rhetoric has effects; it has costs as well. For example, the professional response to creationism, while reasonable on technical, empirical grounds, can also be understood as counterproductive as a means to the end of stemming local efforts to keep fundamentalist dogma out of public school curricula. The ecosystems enacted in and by the debate over cold fusion illustrate the wide-ranging, interpenetrating practices brought to bear on what are taken at first glance to be highly insulated, arcane professional disputes. This suggests that by paying close attention to the intersections of the disparate practices configured in the dispute, we can begin to appreciate the *scientificity* of modern culture and the *rhetoricity* of modern science, a appreciation that rejects what Woolgar called the "exoticism of the other" (1988b, 108). Understood rhetorically, who or what is constituted as "other" is always a contextual production, with equally contextual implications for subsequent configurations.

Of course, understanding science as enacted in a series of local productions makes problematic attempts to craft broad theoretical explanations of demarcation practices. As a consequence, the case studies advanced here remain heuristic, content for now to suggest future directions for subsequent case studies. Hence, while this view provides a theoretical framework, it cannot as yet predict the contours of critical research practices conducted within it.

This need not, of course, be taken as preempting scrutiny, deflecting local critiques by retreating to the comfortable environment of "work in process." Evaluations of the cases here remain the product of a complex scholarly conversation. We should not forget, however, that our conversants do make de facto evaluations of the quality, legitimacy, and accuracy of various conversational offerings. The currency of the conversation, then, must be argumentative (Brockriede 1974). It is this insistence on contextually open-ended argumentation that Billig (1989) sees as the key to avoiding rhetorical conservatism, enacting a meaningful reflexivity, and enabling constructive critique. His position is aptly analogized in the Talmudic parable that concludes, "The law is not in the heavens. The law, indeed all laws, lay in earthly discussions. If heavenly voices wished to join in the debates, they must do better than shake walls or send carobs flying: they must present arguments" (Billig 1987, 29).

Flying Carobs and Arguments: Toward Refinement and Reconstruction

If my central claim—that demarcation practices are both more complex than normally thought and central aspects of scientific practice—is accurate, it is clear that the two case studies detailed here do not exhaust the contexts in which rhetorical demarcation may be played out. Future research might profitably focus on other such contexts. This study has examined contexts in which demarcation lines are drawn for purposes of exclusion from science (creationism) and in which internal demarcation lines are drawn to exclude bad science (cold fusion).

A logical extension of the framework elaborated here would be a context in which consensually accepted lines of external and internal demarcation must be redrawn, such as for purposes of truly interdisciplinary research. A host of literature on the formation of cognitive psychology and other "umbrella disciplines" suggests that such contexts require scientific rhetors to redefine their assumptions regarding the nature of science and the appropriate methods for its investigation in order to facilitate research practice (Abrahamson 1987; Barmark and Wallen 1980; Bechtel 1986, 1987; Darden and Maull 1977).

Future research might also focus on the constraints on the actual closure of demarcation disputes. This study has been content to focus on the competing demarcations and their rhetorical implications. It has consciously avoided any position on the veracity of any particular demarcation. Demarcation disputes generally *do* achieve closure, however. John Campbell's (1986) work on the rhetorical adaptation of Darwinism to the grammar of culture suggests that some degree of consistency with prevailing standards must be maintained by successful demarcation rhetorics. Prelli's (1989b) discussions of rhetorical reasonableness might also provide some preliminary indication of the standards invoked to settle demarcation disputes. On this question, the current study only suggests that some sort of negotiation would take place between rhetorical, structural, and material interests and constraints.

The cold fusion case study seems especially heuristic on this account. Clearly, some sort of closure has been achieved on, at least, the initial claims advanced by Pons and Fleischmann. It is not sufficient, though, to suggest that closure was achieved simply because they were wrong. Their behavior, certainly unorthodox, might well have been deplorable. They might, as some have suggested, have veered perilously close to fraud. Or perhaps, as still others have suggested, they may have been the victims of an oppressive majority view. What does seem clear, though, is that what facticity (presumably the basis for closure) *means* is a contextual production, and it might be produced quite differently as the consitutent audiences increase in number and complexity of goals.

The story told here also suggests the utility of extending a demarcation analysis to the community-defining practices of enterprises that make no claim to scientific status, or would even self-consciously reject such status. Gieryn's (1983b) theoretical explanation of boundary work (which was limited to science in all his subsequent essays) suggested that it is a rhetorical style which could be adapted to the efforts of artisans and metaphysicians as well as scientists. One could reasonably expect to find the rhetorical negotiation of "what makes art art," just as we found the negotiation of "what makes science science." Similarly, the literature on the "professionalization" of various disciplines through identification of their constitutive commitments might have useful things to say to a study in demarcation (e.g., Whitley 1977). Of course, this turns the critical eye toward the university system itself, probing deeply into the origins and maintenance of the value system which underwrites the trinity of research, teaching, and service.

A final implication for future research emerging directly from this analysis involves the representation of demarcation criteria in other discursive contexts (Ormiston and Sassower 1989). As discussed in chapter 5, for example, textbook definitions of science profoundly influence lay perceptions and understandings of the nature of science. Future research might usefully examine the influences on those demarcations, such as the contribution of technical demarcation disputes, political pressure on textbook publishers, and so on (e.g., Nelkin 1977a, 1977b). Similarly, a study of implicit demarcations which are "popularized" in the mass media would have important implications for the social authority of what we have come to recognize as science (e.g., Basalla 1976; Bayertz 1985; Taylor and Condit 1988; Whitley 1985).

A different, but no less important, discursive context involves the articulation of scientific issues in legal forums. While not an explicit focus of the creationism story told here, it might be argued that traditional scientists face a paradox, being damned if they do, and damned if they don't in relation to defining creationism as nonscience. If they avoid communal identifications that would exclude creationism at the cost of appearing authoritarian, then creationist attempts to claim scientific standing are affirmed. If they pursue such identifications, they run the risk of appearing to be an exclusivist elite, concerned less with science than with controlling turf. It is not at all clear that such a paradox can be avoided programmatically. This analysis implies that the consequences of any particular demarcation strategy need bear no relationship to the outcomes presumably intended by the community advancing said strategy. Similarly, judges charged with adjudicating the constitutional standing of any particular religious doctrine masquerading as nontraditional science are bound, by the Constitution and a rightly strong series of precedents, to exclude creationism. The perspective detailed here does not speak against such judgments but suggests that they will necessarily be evaluated along the myriad components of the scientific ecosystem crafted in the controversy.

In the cold fusion and creationism disputes, the normative contours of and hierarchical placement within the ecosystem of the species of science and politics were articulated in the discursive practices of the disputants. Rather than assuming that such outcomes were preordained by the nature of fusion products, by evolutionary processes, or even by the intrinsic nature of science itself, a rhetorical perspective on science as practice views the outcomes as a contextualized symbolic stabilization of the ecosystem.

This act of stabilization belies the traditional presumption of science's difference and distance from other lifeworlds. Mukerji argues, for example, that "the voice of science is authoritative to the extent that it seems objective and above politics even when applied to policy" (1989, 191). Jasanoff engages this connectedness even more directly, arguing that "scientists have an institutional stake in reducing public interaction between science and the administrative process, since these interactions emphasize the indeterminacy and lack of consensus within science, thereby weakening science's claim to cognitive authority" (1987, 224). This attempt to reduce interaction with the public and its formal legislative organs, however, circumvents precisely the valuable contributions science can and should make to public deliberation. As Meyer commented regarding the current environmental crisis, "Scientists have an obligation to inform public policy, and that means being political players. . . . For scientists to pretend to be above the political fray is to consign science to irrelevance in policy making" (1995, B2).

If such is indeed the case, then a compelling implication emerges: the cultural configuration that privileges science over and against politics (and other cultural discourses) is fundamentally discursive. To suggest that any hierarchical configuration of practices is a rhetorical construction entails as well that it is subject to alternative construction—a task to which rhetorical critics should set themselves.

Critical evaluations of what science can and cannot tell or do for (or to) us need not be viewed as somehow anti-intellectual, or antiscience in particular. On the contrary, perhaps it is the artificially inflated conception of scientific expertise that disempowers informed public debate—and is the scientific community's own worst enemy. Recognizing the human, hence self-interested, face of science could, as Collins and Yearley put it, "discourage people from judging science against a criterion of infallibility; since science cannot deliver infallibility, to judge it thus is to risk widespread disillusion" (1992, 309). Such disillusionment is especially problematic in an age when many of the compelling social issues demanding our attention draw, at some level, on technical discourses.

Perhaps if less grandiose cognitive standards and a more reflexive public persona were enacted in the interpenetrating cultural discourses of science, then cynicism and anti-intellectualism would be less likely outcomes of simple disagreement. That which is taken to be an imperfection is meaningful only in relation to expectations of perfection. Rhetorical critics appear especially well suited to contribute to a more meaningful and productive public discourse on the nature of science and culture by privileging humility over hubris—all around.

The importance of this issue extends well beyond preventing scientists from embarrassing themselves, however. Constructive analyses of scientific rhetorics would appear to be a key contributor to the enactment of what Fuller calls "knowledge policy." Abandoning such traditional oppositions as "basic/applied," "internal/external," even "science/society," his knowledge policy enacts "prolescience," a stance that presumes that "knowledge production should proceed only insofar as public involvement is possible" (1993, xviii).

Such a position might strike many as an unwarranted intrusion of the uninformed into matters best left to the more informed, a status Bell (1992) labels "impure science." Such a reaction presumes too hastily that we (and they) can know *in advance* what constitutes the requisite knowledge for categorization as "informed." Beyond this, however, it trades on the too-seldom-examined premise that science is best organized by scientists, brooking only token opposition from and asking only token approval from its constituent audiences. Traweek has argued that "certainly big science requires a big audience. . . . This audience is produced and seduced by scientists. The role of this awed audience for science is not to judge the value of the projects of scientists and engineers, its functions are to approve, fund, and to provide recruits" (1992, 103).

Philosophically, a commitment to a more democratic science policy entails rejecting such condescension on its face. It urges a reconfiguration of this particular way of stabilizing the ecosystem, a way that, at present, privileges the scientific establishment in the short term but bodes ill for it and for the larger culture in the longer term. Understood this way, science is made once again an active, important constituent of the ecosystem, but one that, like any other, must justify its share of the resources to those who make allocation decisions and to those who might, in one way or another, be affected by the potential reallocations and/or the productions funded by them.

This seems especially compelling in an age marked by increasing scarcity of material resources and an exponential growth in demands for them, especially for "megaprojects" such as the Human Genome Project or the ill-fated Superconducting Supercollider. In retrospect, it now seems apparent that the empirical "fates" of the top quark and Higgs's Boson were more dependent on what happened in Washington, D.C., and Austin than on what did or did not happen in (or, more precisely, under) Waxahachie. Martha Krebs, director of the Department of Energy's Office of Energy Research, describing the huge requests for fusion research, conceded that "in the face of some of the political reactions to large scale projects like the SSC, it's really going to take a strong act of persistence and will to sustain the program" (in McDonald 1994, A14). Perhaps, however, it will and should require something more: a strong act of *persuasion*. Ironically, one disgruntled (now unemployed) former technician working on the supercollider seemed to sense this requirement, lamenting that "the work is so far removed from everyday experience, it's hard to see the connection. . . . We don't use a language people can understand" (in E. Lane 1994).

Inadequate understanding is but part of the problem, and enhancing admittedly unacceptable levels of scientific literacy is but part of the solution. Perhaps a larger part of the problem is the assumption that "people," given improved understanding, would be compelled naturally to concede the funds for such projects. Perhaps we would be. Maybe we should be. As Brian Zimmerman has pointed out, however, "The slogan of 'truth for truth's sake' is defunct, simply because science is no longer, and can never again be, the private affair of scientists" (1993, 446). As a consequence, more productive engagements of and judgments on the pressing issues of our time require the crafting of more inclusive, constructive, and democratic rhetorics of science by practitioners, consumers, and critics alike.

BIBLIOGRAPHY

INDEX

BIBLIOGRAPHY

Abir-Am, P. 1987. The biotheoretical gathering, trans-disciplinary authority, and the incipient legitimation of molecular biology in the 1930's: New perspective on the historical sociology of science. *History of Science* 25: 1–69.

Abrahamson, A. 1987. Bridging boundaries versus breaking boundaries: Psycholinguistics in perspective. *Synthese* 72: 355–88.

Affanato, F. 1986. "A survey of biology teachers' opinions about the teaching of evolutionary theory and/or the creation model in the United States in public and private schools." Ph.D. diss., University of Iowa.

Albert, L. 1986. "Scientific" creationism as a pseudoscience. *Creation/Evolution* 18: 25–34.

Albury, W. 1980. Politics and rhetoric in the sociobiology debate. *Social Studies of Science* 10: 519–36.

Alexander, R. 1978. Evolution, creation, and biology teaching. *Science Teacher* 48: 91–103, 107.

Allegretti, D. 1989. Fusion claim puzzles UW profs. *(Madison, Wis.) Capital Times*, 20 April. [In Newsbank Index 40: F9.]

Allison, P. 1979. Experimental parapsychology as a rejected science. In *On the margins of science: The social construction of rejected knowledge*, edited by R. Wallis, 271–92. Sociological Review Monograph 27. Keele: University of Keele.

Altman, L. 1995. Promises of miracles: News releases go where journals fear to tread. *New York Times*, 10 January, B6.

Amsterdamski, S. 1975. *Between experience and metaphysics: Philosophical problems of the evolution of science*. Dordrecht: Reidel.

Anderson, R. 1970. Rhetoric and science journalism. *Quarterly Journal of Speech* 56: 358–68.

Anderson, V. 1983. Scientific creationism and its critique of evolution. In *Evolution versus creationism*, edited by J. Zetterberg, 235–48. Phoenix: Oryx Press.

Anderson, W. 1989. Scientific nomenclature and revolutionary rhetoric. *Rhetorica* 7: 45–53.

Applegate, J. 1989. Firms rush to find out about fusion experiments. *Los Angeles Times*, 13 April. [In Newsbank Index 40: F2–F3.]

Aristotle. 1991. *On rhetoric: A theory of civic discourse*. Translated by G. Kennedy. Oxford: Oxford University Press.

Armstrong, H. 1980. Evolutionistic defense against thermodynamics disproved. *Creation Research Society Quarterly* 16: 226–27.

Aronowitz, S. 1988a. The production of scientific knowledge: Science, ideology, and Marxism. In *Marxism and the interpretation of culture*, edited by C. Nelson and L. Grossberg, 519–37. Urbana: University of Illinois Press.

Aronowitz, S. 1988b. *Science as power: Discourse and ideology in modern society.* Minneapolis: University of Minnesota Press.

Arons, A. 1983. Achieving wider scientific literacy. *Daedalus* 112: 91–122.

Ashmore, M. 1989. *The reflexive thesis: Wrighting sociology of scientific knowledge.* Chicago: University of Chicago Press.

Asimov, I. 1984. The "threat" of creationism. In *Science and creationism,* edited by A. Montagu, 182–93. New York: Oxford University Press.

Austin, S., and J. Morris. 1986. Tight folds and clastic dikes as evidence for rapid deposition and deformation of two very thick stratigraphic sequences. In *Proceedings of the First International Conference on Creationism,* edited by R. Walsh, C. Brooks, and R. Crowell, 3–16. Pittsburgh: Creation Science Fellowship.

Awbrey, F., and W. Thwaites. 1984. Introduction to *Evolutionists confront creationists,* edited by F. Awbrey and W. Thwaites, 5–7. San Francisco: Pacific Division, American Association for the Advancement of Science.

Babcock, B. 1980. Reflexivity: Definitions and discriminations. *Semiotica* 30: 1–14.

Bacon, F. 1900. *The advancement of learning and Novum organum.* Rev. ed. New York: Willey Books.

Bakken, G. 1990. *Creation or evolution.* Berkeley: National Center for Science Education.

Balanced treatment for creation science and evolution science, Act 590. 1981. Arkansas Stat. Ann. 580–1663 Supp.

Ballantine, L. 1982. Letter. *Physics Today* 35: 11.

Balthrop, V. 1989. W(h)ither the public sphere? An optimistic reading. In *Spheres of Argument: Proceedings of the Sixth SCA/AFA Conference on Argumentation,* edited by B. Gronbeck, 20–26. Annandale, Va.: Speech Communication Association.

Bantz, C. 1981. Public arguing in the regulation of health and safety. *Western Journal of Speech Communication* 65: 71–87.

Barber, B. 1952. *Science and the social order.* New York: Macmillan.

Barber, B. 1987. Trust in science. *Minerva* 25: 123–34.

Barinaga, M. 1989. California backs evolution education. *Science* 246: 881.

Barker, E. 1979. In the beginning: The battle of creationist science against evolutionism. In *On the margins of science: The social construction of rejected knowledge,* edited by R. Wallis, 179–200. Sociological Review Monograph 27. Keele: University of Keele.

Barmark, J., and G. Wallen. 1980. The development of an interdisciplinary project. In *The social process of scientific investigation,* edited by K. Knorr, R. Krohn, and R. Whitley, 221–35. Dordrecht: Reidel.

Barnes, B. 1974. *Scientific knowledge and sociological theory.* London: Routledge and Kegan Paul.

Barnes, B. 1977. *Interests and the growth of knowledge.* London: Routledge and Kegan Paul.

Barnes, B. 1981. On the conventional character of knowledge and cognition. *Philosophy of the Social Sciences* 11: 303–33.

Barnes, B. 1982. On the extensions of concepts and the growth of knowledge. *Sociological Review* 30: 23–44.

Barnes, B. 1983. *T. S. Kuhn and social science.* New York: Columbia University Press.

Barnes, B. 1984. Problems of intelligibility and paradigm instances. In *Scientific rationality: The sociological turn,* edited by J. Brown, 113–26. Boston: Reidel.

Barnes, B. 1985. A case of amnesia? *Social Studies of Science* 15: 175–76.

Barnes, B., and D. Bloor. 1982. Relativism, rationalism, and the sociology of knowledge. In *Rationality and relativism,* edited by M. Hollis and S. Lukes, 1–20. Cambridge: MIT Press.

Barnes, B., and R. Dolby. 1970. The scientific ethos: A deviant viewpoint. *Archives of the European Journal of Sociology* 11: 3–25.

Barnes, B., and D. Edge, eds. 1982. *Science in context: Readings in the sociology of science.* Cambridge: MIT Press.

Barnes, B., and D. MacKenzie. 1979. The role of interests in scientific change. In *On the margins of science: The social construction of rejected knowledge,* edited by R. Wallis, 49–66. Sociological Review Monograph 27. Keele: University of Keele.

Barnes, B., and S. Shapin. 1977. Where is the edge of objectivity? *British Journal of the History of Science* 10: 61–66.

Bartley, W. 1968a. Reply. In *Problems in the philosophy of science,* edited by I. Lakatos and A. Musgrave, 102–19. Amsterdam: North Holland.

Bartley, W. 1968b. Theories of demarcation between science and metaphysics. In *Problems in the philosophy of science,* edited by I. Lakatos and A. Musgrave, 40–64. Amsterdam: North Holland.

Basalla, G. 1976. Pop science: The depiction of science in popular culture. In *Science and its public: The changing relationship,* edited by G. Holton and W. Blanpied, 261–78. Dordrecht: Reidel.

Bastide, F. 1992. A night with Saturn. Translated by G. Myers. *Science, Technology, and Human Values* 17: 259–81.

Bauer, H. 1983. Velikovsky and the Loch Ness Monster: Attempts at demarcation in two controversies. In *Working papers on the demarcation of science and pseudoscience,* edited by R. Laudan, 87–106. Blacksburg: Virginia Tech Center for the Study of Science in Society.

Baumgardner, J. 1986. Numerical simulation of the large-scale tectonic changes accompanying the flood. In *Proceedings of the First International Conference on Creationism,* edited by R. Walsh, C. Brooks, and R. Crowell, 17–30. Pittsburgh: Creation Science Fellowship.

Bayertz, K. 1985. Spreading the spirit of science: Social determinants of the popularisation of science in nineteenth-century Germany. In *Expository science: Forms and functions of popularisation,* edited by T. Shinn and R. Whitley, 209–28. Dordrecht: Reidel.

Bazerman, C. 1981. What written knowledge does: Three examples of academic discourse. *Philosophy of the Social Sciences* 11: 361–87.

Bazerman, C. 1984. The writing of scientific non-fiction. *Pre/Text* 6: 9–29.

Bazerman, C. 1985. Physicists reading physics: Schema-laden purposes and purpose-laden schema. *Written Communication* 2: 3–23.

Bazerman, C. 1987. Codifying the social scientific style: The APA publication manual as a behaviorist rhetoric. In *The rhetoric of the human sciences: Language and argument in scholarship and public affairs,* edited by J. Nelson, A. Megill, and D. McCloskey, 125–44. Madison: University of Wisconsin Press.

Bazerman, C. 1988. *Shaping written knowledge: The genre and activity of the experimental article in science.* Madison: University of Wisconsin Press.

Beall, H., and L. Berka. 1990. Report on the WPI-NEACT Conference: Perceptions of chemistry. *Journal of Chemical Education* 67: 103–4.

Bechtel, W. 1986. The nature of scientific integration. In *Science and philosophy: Integrating scientific disciplines,* edited by W. Bechtel, 3–52. Dordrecht: Martinus Nijhoff.

Bechtel, W. 1987. Psycholinguistics as a case of cross-disciplinary research. *Synthese* 72: 293–311.

Begley, S. 1989. The race for fusion. *Newsweek,* 8 May, 49–55.

Bell, R. 1992. *Impure science: Fraud, compromise, and political influence in scientific research.* New York: Wiley.

Ben-Chaim, M. 1992. The empiric experience and the practice of autonomy. *Studies in the History and Philosophy of Science* 23: 533–55.

Ben-David, J. 1958. The professional role of the physician in bureaucratized medicine: A study in role conflict. *Human Relations* 11: 255–74.

Ben-David, J. 1981. Sociology of scientific knowledge. In *The state of sociology,* edited by J. Short, 12–65. Beverly Hills: Sage.

Ben-David, J. 1984. *The scientist's role in society: A comparative study.* Chicago: University of Chicago Press.

Bennetta, W. 1986. *Crusade of the credulous.* San Francisco: California Academy of Sciences Press.

Bergman, J. 1982. Does academic freedom apply to both secular humanists and Christians? In *Creation: The cutting edge,* edited by H. Morris and D. Rohrer, 24–32. San Diego: Creation-Life.

Berkenkotter, C., T. Huckin, and J. Ackerman. 1991. Social context and socially constructed texts: The initiation of a graduate student into a writing research community. In *Textual dynamics of the professions,* edited by C. Bazerman and J. Paradis, 191–215. Madison: University of Wisconsin Press.

Bernabo, L., and C. Condit. 1990. Two stories of the Scopes trial: Legal and journalistic articulations of the legitimacy of science and religion. In *Popular trials: Rhetoric, mass media, and the law,* edited by R. Hariman, 55–85. Tuscaloosa: University of Alabama Press.

Bernick, B., and J. Jacobsen-Wells. 1989. Fusion: BYU's research went to peers, not public. Salt Lake City *Deseret News,* 28 March. [In Newsbank Index 40: C14.]

Bernstein, R. 1982. *Beyond objectivism and relativism.* Philadelphia: University of Pennsylvania Press.

Beyerstein, D. 1990. Dealing with creationist rhetoric. *Creation/Evolution* 27: 20–28.

Bhaskar, R. 1978. *A realist theory of science.* London: Harvester Press.

Biesecker, B. 1989. Recalculating the relation of the public and technical spheres. In *Spheres of argument: Proceedings of the Sixth SCA/AFA Conference on Argumentation,* edited by B. Gronbeck, 66–70. Annandale, Va.: Speech Communication Association.

Billig, M. 1987. *Arguing and thinking: A rhetorical approach to social psychology.* Cambridge: Cambridge University Press.

Billig, M. 1989. Conservatism and the rhetoric of rhetoric. *Economy and Society* 18: 132–48.

Billig, M. 1991. *Ideology and opinions: Studies in rhetorical psychology.* London: Sage.

Bird, W. 1978. Freedom of religion and science instruction in public schools. *Yale Law Journal* 87: 515–70.

Bishop, J. 1989a. Scientist sticks to claim of sustained test tube fusion. *Wall Street Journal,* 27 March, B3.

Bishop, J. 1989b. Second fusion discovery comes to light. *Wall Street Journal,* 29 March, B4.

Bishop, J., and K. Wells. 1989. Taming H-bombs? Two scientists claim breakthrough in quest for fusion energy. *Wall Street Journal,* 24 March, 1, 7.

Bjerklie, D. 1989. Fusion illusion. *Time,* 8 May, 72–77.

Black, E. 1968. *Rhetorical criticism: A study in method.* Madison: University of Wisconsin Press.

Black, M. 1934. The principle of verifiability. *Analysis* 2: 1–6.

Bliss, R. 1983. A two-model approach to origins: A curriculum imperative. In *Evolution versus creationism,* edited by J. Zetterber, 192–97. Phoenix: Oryx Press.

Bloor, D. 1973. Wittgenstein and Mannheim on the sociology of mathematics. *Studies in the History and Philosophy of Science* 4: 173–91.

Bloor, D. 1976. *Knowledge and social imagery.* Boston: Routledge and Kegan Paul.

Bloor, D. 1981. The strengths of the strong programme. *Philosophy of the Social Sciences* 11: 199–213.

Bloor, D. 1984. The sociology of reasons; or, Why "epistemic factors" are really "social factors." In *Scientific rationality: The sociological turn,* edited by J. Brown, 295–324. Boston: Reidel.

Bloor, D. 1989. Professor Campbell on models of language learning and the sociology of science: A reply. In *The cognitive turn: Sociological and psychological perspectives on science,* edited by S. Fuller, M. De Mey, T. Shinn, and S. Woolgar, 159–66. Dordrecht: Kluwer.

Boardman, W., R. Koontz, and H. Morris. 1973. *Science and creation.* San Diego: Creation-Life.

Bockris, J., and D. Hodko. 1990. Is there evidence for cold fusion? *Chemistry and Industry,* 5 November, 688–92.

Bohme, G. 1977. Cognitive norms, knowledge-interests, and the constitution of the scientific object: A case study in the functioning of rules for experimentation. In *The social production of scientific knowledge,* edited by E. Mendelsohn, P. Weingart, and R. Whitley, 129–42. Dordrecht: Reidel.

Bohme, G. 1979. Alternatives in science—alternatives to science. In *Counter movements in the sciences*, edited by H. Nowotny and H. Rose, 105–25. Dordrecht: Reidel.

Bohme, G., and N. Stehr. 1986. *The knowledge society*. Boston: Reidel.

Bokeno, R. 1987. The rhetorical understanding of science: An explication and critical commentary. *Southern Speech Communication Journal* 52: 285–311.

Boreman, B., and D. Kim. 1981. Governing the "republic of science": An analysis of NSF officials' attitudes toward managed science. *Polity* 14: 183–204.

Boulding, K. 1984. Toward an evolutionary theology. In *Science and creationism*, edited by A. Montagu, 142–58. New York: Oxford University Press.

Bowden, G. 1985. The social construction of validity in estimates of U.S. crude oil reserves. *Social Studies of Science* 15: 2.

Bowden, M. 1981. *Ape-men: Fact or fallacy?* Bromley, Kent: Sovereign.

Bowler, P. 1989. *Evolution: The history of an idea*. Berkeley: University of California Press.

Bozeman, T. 1977. *Protestants in an age of science: The Baconian ideal and antebellum American religious thought*. Chapel Hill: University of North Carolina Press.

Broad, W. 1981. Creationists limit scope of California case. *Science* 211: 1331–32.

Broad, W. 1989a. At conference on cold fusion, verdict is negative. *New York Times*, 30 May, 19, 22.

Broad, W. 1989b. Conference on fusion told of failure. *New York Times*, 24 May, 12.

Broad, W. 1989c. A frenzy over fusion in hundreds of labs. *New York Times*, 18 April, 25.

Broad, W. 1989d. Fusion in a jar: Recklessness and brilliance. *New York Times*, 9 May, 23.

Broad, W. 1989e. Fusion researchers seek $25 million from U.S. *New York Times*, 26 April, 10.

Broad, W. 1989f. House panel hears debate of scientists' fusion claim. *New York Times*, 27 April, 13.

Broad, W. 1989g. New form of "hot" fusion achieved. *New York Times*, 12 September, 22.

Broad, W. 1989h. Panel rejects fusion claim, urging no federal spending. *New York Times*, 13 July, 1, 10.

Broad, W. 1989i. Scientists dispute results of experiments with cold fusion. *New York Times*, 25 May, 13.

Broad, W. 1991. There may still be something scientific about cold fusion. *New York Times*, 14 April, E4.

Broad, W., and N. Wade. 1982. *Betrayers of the truth*. New York: Simon and Schuster.

Brockriede, W. 1974. Rhetorical criticism as argument. *Quarterly Journal of Speech* 60: 165–74.

Brown, G. 1993. Technology's dark side. *Chronicle of Higher Education*, 30 June, B1–B2.

Brown, M. 1989. Cold fusion, hot products. Salt Lake City *Deseret News*, 21 May. [In Newsbank Index 53: A4–A5.]

Brown, W., and R. Crable. 1973. Industry, mass magazines, and the ecology issue. *Quarterly Journal of Speech* 59: 259–72.

Browne, M. 1989a. Fusion claim is greeted with scorn by physicists. *New York Times,* 3 May, 1, 13.
Browne, M. 1989b. Physicists challenge cold fusion claims. *New York Times,* 2 May, 17, 22.
Brush, S. 1981. Creationism/evolution: The case against equal time. *Science Teacher* 48: 29–37.
Brush, S. 1983. Ghosts from the nineteenth century: Creationist arguments for a young earth. In *Scientists confront creationism,* edited by L. Godfrey, 49–84. New York: Norton.
Brush, S. 1986. Skepticism: Another alternative to science or belief. In *Science and creation,* edited by R. Hanson, 160–73. New York: Macmillan.
Bryant, D. 1965. Rhetoric: Its functions and scope. In *The province of rhetoric,* edited by J. Schwartz and J. Rycenga, 3–36. New York: Ronald Press.
Buderi, R. 1989. No more creationism. *Nature* 341: 561.
Bunders, J., and R. Whitley. 1985. Popularisation within the sciences: The purposes and consequences of inter-specialist communication. In *Expository science: Forms and functions of popularisation,* edited by T. Shinn and R. Whitley, 61–78. Dordrecht: Reidel.
Bunge, M. 1991. A critical examination of the new sociology of science. *Philosophy of the Social Sciences* 21: 524–60.
Burke, K. 1962. *A grammar of motives and a rhetoric of motives.* Cleveland: World Meridian.
Bytwerk, R. 1970. The SST controversy: A case study in the functioning of rules for experimentation. *Central States Speech Journal* 30: 187–98.
Calais, R., and G. Duffett. 1987. A theory for the birds? *Creation Research Society Quarterly* 24: 183–85.
Cale, B. 1972. Darwin and the concept of a struggle for existence: A study in the extra-scientific origins of scientific ideas. *Isis* 63: 321–44.
Cambrosio, A., and P. Keating. 1983. The disciplinary stake: The case of chronobiology. *Social Studies of Science* 13: 323–53.
Campbell, D. 1989. Models of language learning and their implications for social constructionist analyses of scientific belief. In *The cognitive turn: Sociological and psychological perspectives on science,* edited by S. Fuller, M. De Mey, T. Shinn, and S. Woolgar, 153–58. Dordrecht: Kluwer.
Campbell, J. 1974. Charles Darwin and the crisis of ecology: A rhetorical perspective. *Quarterly Journal of Speech* 60: 442–49.
Campbell, J. 1975. The polemical Mr. Darwin. *Quarterly Journal of Speech* 61: 375–90.
Campbell, J. 1983. Creationism: The argument that time forgot. In *Argument in transition: Proceedings of the Third SCA/AFA Summer Conference on Argumentation,* edited by D. Zaresky, M. Sillars, and J. Rhodes, 423–40. Annandale, Va.: Speech Communication Association.
Campbell, J. 1986. Scientific revolution and the grammar of culture. *Quarterly Journal of Speech* 72: 351–76.
Campbell, J. 1987. Charles Darwin: Rhetorician of science. In *The rhetoric of the human sciences: Language and argument in scholarship and public affairs,* edited

by J. Nelson, A. Megill, and D. McCloskey, 69–86. Madison: University of Wisconsin Press.

Campbell, J. 1989. The invisible rhetorician: Charles Darwin's "third party" strategy. *Rhetorica* 7: 55–85.

Campbell, J. 1993. Reply to Gaonkar and Fuller. *Southern Communication Journal* 58: 312–18.

Campbell, P. 1975. The personae of scientific discourse. *Quarterly Journal of Speech* 61: 391–405.

Cantor, G. 1975. The Edinburgh phrenology debate: 1803–1828. *Annals of Science* 32: 195–218.

Carey, J. 1989. Fusion in a bottle: Miracle or mistake? *Business Week,* 8 May, 100–110.

Carleton, W. 1978. What is rhetorical knowledge? A response to Farrell—and more. *Quarterly Journal of Speech* 64: 313–28.

Carlston, D. 1987. Turning psychology on itself: The rhetoric of psychology and the psychology of rhetoric. In *The rhetoric of the human sciences: Language and argument in scholarship and public affairs,* edited by J. Nelson, A. Megill, and D. McCloskey, 145–62. Madison: University of Wisconsin Press.

Carnap, R. [1932] 1959. Verberwindung der metaphysik durch logische analyse der sprache. English translation in *Logical positivism,* edited by A. Ayer, 60–81. Glencoe, Ill.: Free Press.

Carnap, R. 1936. Testability and meaning. *Philosophy of Science* 3: 419–71.

Carnap, R. 1937. *The logical syntax of language.* London: Routledge and Kegan Paul.

Carnap, R. 1967. *The logical structure of the world.* Berkeley: University of California Press.

Carr, W. 1985. Philosophy, values, and educational science. *Journal of Curriculum Studies* 17: 119–32.

Casti, J. 1989. *Paradigms lost: Images of man in the mirror of science.* New York: William Morrow.

Cavanaugh, M. 1985. Scientific creationism and rationality. *Nature* 315: 185–89.

Cavanaugh, M. 1987. One-eyed social movements: Rethinking issues in rationality and society. *Philosophy of the Social Sciences* 17: 147–72.

Ceccarelli, L. 1994. A masterpiece in a new genre: The rhetorical negotiation of two audiences in Schrodinger's "What is life?" *Technical Communication Quarterly* 3: 7–20.

Chaikowsky, J. 1979. The ponderings of an optimistic evolutionist. *Creation Research Society Quarterly* 16: 174.

Chalmers, A. 1986. The Galileo that Feyerabend missed: An improved case against method. In *The politics and rhetoric of scientific method,* edited by J. Schuster and R. Yeo, 1–31. Dordrecht: Reidel.

Chandler, D. 1989a. But at least part of the claims are gaining more credence. *Boston Globe,* 17 April. [In Newsbank Index 40: D6–D9.]

Chandler, D. 1989b. Emphasis again on trying to mimic sun's core. *Boston Globe,* 5 June. [In Newsbank Index 65: G2–G3.]

Chandler, D. 1989c. MIT scientist offers theory for room temperature fusion. *Boston Globe,* 15 April. [In Newsbank Index 40: D5.]
Chapline, G. 1989. Cold water on cold fusion. *Chicago Tribune,* 25 July, 15.
Charland, M. 1987. Constitutive rhetoric: The case of the Peuples Québécois. *Quarterly Journal of Speech* 73: 133–50.
Charland, W. 1989. Academic capitalism: Did greed sell out cold fusion breakthrough? *Champaign-Urbana News Gazette,* 23 July, B1.
Charlesworth, M. 1986. *Science, non-science, and pseudoscience.* Burwood, Victoria: Deakin University Press.
Chatterjee, L. 1989. More on cold fusion. *Nature* 342: 232.
Chen, X., and P. Barker. 1992. Cognitive appraisal and power: David Brewster, Henry Brougham, and the tactics of the emission-undulatory controversy during the early 1850s. *Studies in the History and Philosophy of Science* 23: 75–101.
Chomsky, N. 1982. *Towards a new cold war.* New York: Pantheon.
Christian, P. 1989. Chemists take on physicists in cold war. *Provo (Utah) Daily Herald,* 5 May. [In Newsbank Index 52: F12–F13.]
Chubin, D. 1990. Scientific malpractice and the contemporary politics of knowledge. In *Theories of science in society,* edited by S. Cozzens and T. Gieryn, 144–63. Bloomington: Indiana University Press.
Chubin, D., and E. Hackett. 1990. *Peerless science: Peer review and U.S. science policy.* Albany: State University of New York Press.
Ciotti, P. 1989. Fear of fusion: What if it works? *Los Angeles Times,* 19 April. [In Newsbank Index 40: G9.]
Clark, R. 1991. What ever happened to cold fusion? *Journal of Chemical Education* 68: 277–79.
Cloitre, M., and Shinn, T. 1985. Expository practice: Social, cognitive, and epistemological linkage. In *Expository science: Forms and functions of popularisation,* edited by T. Shinn and R. Whitley, 31–60. Dordrecht: Reidel.
Close, F. 1990. *Too hot to handle: The race for cold fusion.* Princeton, N.J.: Princeton University Press.
Cloud, D. 1989. Utah's nuclear fusion scheme may create fission on Hill. *Congressional Quarterly Weekly Report* 47: 960.
Cohen, J., and J. Davies. 1989a. The cold fusion family. *Nature* 338: 705–7.
Cohen, J., and J. Davies. 1989b. Is cold fusion hot? *Nature* 342: 487–88.
Cohen, L. 1980. Some historical remarks on the Baconian conception of probability. *Journal of the History of Ideas* 41: 219–31.
Cold (con)fusion. 1989. *Nature* 338: 361–62.
Cold fusion scientist quits Utah position. 1991. *Chicago Tribune,* 11 January, 4.
Cold fusion's discoverers go far from the maddening crowd. 1992. *Business Week* 3290: 89.
Cole, H., and E. Scott. 1982. Creation science and scientific research. *Phi Beta Kappan* 63: 557–58.
Cole, J. 1981. Misquoted scientists respond. *Creation/Evolution* 6: 34–44.
Cole, J. 1984. Review of *Evolution: The fossils say no!* by D. Gish. In *Reviews of*

thirty-one creationist books, edited by S. Weinberg, 24–25. Syosset, N.Y.: National Center for Science Education.

Cole, J. 1988. Creationism and the New Right agenda: An opinion survey. *Creation/Evolution* 22: 6–12.

Cole, J., and S. Cole. 1973. *Social stratification in science.* Chicago: University of Chicago Press.

Cole, J., and H. Zuckerman. 1975. The emergence of a scientific specialty: The self-exemplifying case of the sociology of science. In *The idea of social structure: Papers in honor of Robert J. Merton,* edited by L. Coser, 139–74. New York: Harcourt, Brace, Jovanovich.

Cole, S. 1970. Professional standing and the reception of scientific discoveries. *American Journal of Sociology* 76: 286–306.

Collins, H. 1975. The seven sexes: A study in the sociology of a phenomenon or the replication of experiments in physics. *Sociology* 9: 205–24.

Collins, H. 1981a. The place of the core set in modern science: Social contingency with methodological propriety in science. *History of Science* 19: 6–19.

Collins, H. 1981b. Stages in the empirical programme of relativism. *Social Studies of Science* 11: 3–10.

Collins, H. 1981c. What is TRASP?: The radical programme as a methodological imperative. *Philosophy of the Social Sciences* 11: 215–24.

Collins, H. 1982a. Knowledge, norms, and rules in the sociology of science. Social Studies of Science 12: 299–309.

Collins, H. 1982b. The replication of experiments in physics. In *Science in contexts: Readings in the sociology of science,* edited by B. Barnes and D. Edge, 94–116. Cambridge: MIT Press.

Collins, H. 1982c. Special relativism: The natural attitude. *Social Studies of Science* 12: 139–43.

Collins, H. 1983a. An empirical relativist programme in the sociology of scientific knowledge. In *Science observed: Perspectives on the social study of science,* edited by K. Knorr-Cetina and M. Mulkay, 85–114 London: Sage.

Collins, H. 1983b. The sociology of scientific knowledge: Studies of contemporary science. *Annual Reviews of Sociology* 9: 265–85.

Collins, H. 1984. When do scientists prefer to vary their experiments? *Studies in the History and Philosophy of Science* 15: 169–74.

Collins, H. 1985. *Changing order: Replication and induction in scientific practice.* Beverly Hills: Sage.

Collins, H., and T. Pinch. 1979. The construction of the paranormal: Nothing unscientific is happening. In *On the margins of science: The social construction of rejected knowledge,* edited by R. Wallis, 237–70. Sociological Review Monograph 27. Keele: University of Keele.

Collins, H., and S. Yearley. 1992. Epistemological chicken. In *Science as practice and culture,* edited by A. Pickering, 301–26. Chicago: University of Chicago Press.

Collins, R., and S. Restivo. 1983. Development, diversity, and conflict in the sociology of science. *Sociological Quarterly* 24: 185–200.

Colvin, P. 1977. Ontological and epistemological commitments and social relations in the sciences: The case of the arithmomorphic system of scientific production. In *The social production of scientific knowledge,* edited by E. Mendelsohn, P. Weingart, and R. Whitley, 103–28. Dordrecht: Reidel.
Condit, C. 1989. The rhetorical limits of polysemy. *Critical Studies in Mass Communication* 6: 103–22.
Condit, C. 1990. The birth of understanding: Chaste science and the harlot of the arts. *Communication Monographs* 57: 323–27.
Condit-Railsback, C. 1983. Beyond rhetorical relativism: A structural material model of truth and objective reality. *Quarterly Journal of Speech* 69: 351–63.
"[Con]fusion in a Bottle." 1991. Videorecording of *NOVA,* produced by WBGH.
Cooke, R. 1989. In a fever over cold fusion. *Newsday,* 25 April. [In Newsbank Index 41: D12–E3.]
Cookson, C. 1989. Nuclear fusion in a test tube. *Financial Times,* 23 March, 26.
Cooter, R. 1976. Phrenology: The provocation of progress. *History of Science* 14: 211–34.
Cooter, R. 1980. Deploying pseudo-science: Then and now. In *Science, pseudoscience, and society,* edited by M. Hanen, M. Osler, and R. Weyant, 237–61. Waterloo, Ont.: Wilfred Laurier University Press.
Corry, L. 1993. Kuhnian issues, scientific revolutions, and the history of mathematics. *Studies in the History and Philosophy of Science* 24: 95–117.
Cozzens, S. 1986. Funding and knowledge growth: Editor's introduction. *Social Studies of Science* 16: 9–21.
Cozzens, S. 1990. Autonomy and power in science. In *Theories of science in society,* edited by S. Cozzens and T. Gieryn, 164–84. Bloomington: Indiana University Press.
Cozzens, S., and T. Gieryn. 1990. Introduction: Putting science back into society. In *Theories of science in society,* edited by S. Cozzens and T. Gieryn, 1–14. Bloomington: Indiana University Press.
Cracraft, J. 1983. The scientific response to creationism. In *Creationism, science, and the law,* edited by M. LaFollette, 104–13. Cambridge: MIT Press.
Cracraft, J. 1984. The significance of the data of systematics and paleontology for the evolution-creationism controversy. In *Evolutionists confront creationists,* edited by F. Awbrey and W. Thwaites, 189–205. San Francisco: Pacific Division, American Association for the Advancement of Science.
Crane, D. 1972. *Invisible colleges.* Chicago: University of Chicago Press.
Crane, D. 1980. An exploratory study of Kuhnian paradigms in theoretical high energy physics. *Social Studies of Science* 10: 23–54.
Crawford, G. 1983. Science as an apologetic: A tool for biblical literalists. In *Creationism, science, and the law,* edited by M. LaFollette, 104–13. Cambridge: MIT Press.
Crawford, M. 1989. Budget squeeze causes fission in fusion labs. *Science* 244: 423.
Crease, R., and C. Mann. 1986. *The second creation: Makers of the revolution in twentieth-century physics.* New York: Macmillan.
Creationist schools. 1992. *Impact* 225: i–iv.

Crombie, A. 1980. Science and the arts in the Renaissance: The search for truth and certainty. *History of Science* 28: 233–46.

Cuffey, R. 1984. Paleontologic evidence and organic evolution. In *Science and creationism,* edited by A. Montagu, 255–81. New York: Oxford University Press.

Culliton, B. 1978. Science's restive public. *Daedulus* 107: 147–56.

Cunningham, A. 1988. Getting the game right: Some plain words on the identity and invention of science. *Studies in the History and Philosophy of Science* 19: 365–89.

Cushing, J. 1990. Is scientific methodology interestingly atemporal? *British Journal for the Philosophy of Science* 41: 177–94.

Czubaroff, J. 1973. Intellectual respectability: A rhetorical problem. *Quarterly Journal of Speech* 59: 155–64.

Czubaroff, J. 1989. The deliberative character of strategic scientific debate. In *Rhetoric in the human sciences,* edited by H. Simons, 28–47. Beverly Hills: Sage.

Dalrymple, G. 1984. How old is the earth? A reply to "scientific" creationism. In *Evolutionists confront creationists,* edited by F. Awbrey and W. Thwaites, 66–131. San Francisco: Pacific Division, American Association for the Advancement of Science.

Darden, L., and N. Maull. 1977. Interfield theories. *Philosophy of Science* 44: 43–64.

Daston, L. 1978. British responses to psycho-physiology, 1860–1900. *Isis* 69: 192–208.

Dauber, C. 1989. Fusion criticism: A call to criticism. In *Spheres of argument: Proceedings of the Sixth SCA/AFA Conference on Argumentation,* edited by B. Gronbeck, 33–36. Annandale, Va.: Speech Communication Association.

Davenport, E. 1987. The new politics of knowledge: Rorty's pragmatism and the rhetoric of the human sciences. *Philosophy of the Social Sciences* 17: 377–94.

Davis, B. 1986. Molecular genetics and the falsifiability of evolution. In *The kaleidoscope of science,* edited by E. Ullman-Margalit, 95–110. Dordrecht: Reidel.

Davis, H. 1982. What is a scientific theory? *Physics Today* 35: 28.

Dawson, J. 1989. Cold fusion debate. *Minneapolis Star and Tribune,* 7 May. [In Newsbank Index 52: E14–F1.]

Deason, G. 1986. Reformation theology and the mechanistic conception of life. In *God and nature: Historical essays on the encounter between science and Christianity,* edited by D. Lindberg and R. Numbers, 167–91. Berkeley: University of California Press.

De La Bruheze, A. 1992. Radiological weapons and radioactive waste in the United States: Insiders' and outsiders' views, 1941–55. *British Journal of the History of Science* 25: 207–27.

Delia, J. 1977. Constructivism and the study of human communication. *Quarterly Journal of Speech* 63: 66–83.

Delia, J., and L. Grossberg. 1977. Interpretation and evidence. *Western Journal of Speech Communication* 41: 32–42.

Derber, C., W. Schwartz, and Y. Magrass. 1990. *Power in the highest degree: Professionals and the rise of a new Mandarin order*. New York: Oxford University Press.

Detjen, J. 1989a. Scientists fearing fallout from fusion controversy. *Philadelphia Inquirer,* 30 April. [In Newsbank Index 41: A8–A9.]

Detjen, J. 1989b. Scientists stir debate on fusion. *Philadelphia Inquirer,* 2 April. [In Newsbank Index 40: D9–D10.]

Devries, G., and H. Harbers. 1985. Attuning science to culture: Scientific and popular discussion in Dutch sociology of education, 1960–1980. In *Expository science: Forms and functions of popularisation,* edited by T. Shinn and R. Whitley, 103–18. Dordrecht: Reidel.

DeYoung, D. 1992. The plasma universe. *Impact* 228: i–iv.

Dickinson, J., L. Jensen, S, Langford, R. Ryan, and E. Garcia. 1990. Fracto-emission from deuterated titanium: Supporting evidence for a fracto-fusion mechanism. *Journal of Materials Research* 5: 109–22.

Dickson, D. 1988. *The new politics of science.* Chicago: University of Chicago Press.

Dobzhansky, T. 1983. Nothing in biology makes sense except in the light of evolution. In *Evolution versus creationism,* edited by J. Zetterberg, 18–28. Phoenix: Oryx Press.

Does creation deserve equal time? 1981. *Nature* 291: 271–72.

Dolby, R. 1975. What can we usefully learn from the Velikovsky affair? *Social Studies of Science* 5: 165–75.

Dolby, R. 1979. Reflections on deviant science. In *On the margins of science: The social construction of rejected knowledge,* edited by R. Wallis, 9–48. Sociological Review Monograph 27. Keele: University of Keele.

Dolby, R. 1982. On the autonomy of pure science: The construction and maintenance of barriers between scientific establishments and popular culture. In *Scientific establishments and hierarchies,* edited by N. Elias, H. Martins, and R. Whitley, 242–85. Dordrecht: Reidel.

Donahue, G., P. Tichenor, and C. Olien. 1973. Mass media functions, knowledge, and social control. *Journalism Quarterly* 50: 652–59.

Doolittle, R. 1983. Probability and the origin of life. In *Scientists confront creationism,* edited by L. Godfrey, 85–98. New York: Norton.

Douglas, J. 1969. The rhetoric of science and the origins of statistical social thought: The case of Durkheim's "suicide." In *The phenomenon of sociology,* edited by E. Tiryakian, 44–57. New York: Appleton-Century Crofts.

Duffy, B. 1987. Fundamentalism, relativism, and commitment. *Communication Education* 36: 403–9.

Duncan, H. 1968. *Communication and social order.* New Brunswick, N.J.: Transaction Press.

Dunwoody, S., and M. Ryan. 1985. Scientific barriers to the popularization of science in the mass media. *Journal of Communication* 35: 1–12.

Dupree, A. 1986. Christianity and the scientific community in the age of Darwin. In *God and nature: Historical essays on the encounter between science and Christianity,* edited by D. Lindberg and R. Numbers, 351–68. Berkeley: University of California Press.

Dye, L., and T. Maugh. 1989a. Fusion claim sparks rush to duplicate experiment. *Los Angeles Times,* 25 March. [In Newsbank Index 28: E13–E15.]

Dye, L., and T. Maugh. 1989b. Pair proclaim nuclear fusion breakthrough. *Los Angeles Times*, 24 March. [In Newsbank Index 28: F3–F5.]
Eastwood, B. 1984. Descartes on refraction: Scientific vs. rhetorical method. *Isis* 75: 481–50.
Eckberg, D., and A. Nesterenko. 1985. For and against evolution: Religion, social class, and the symbolic universe. *Social Science Journal* 22: 1–17.
Eckert, M. 1987. Propaganda in science: Sommerfeld and the spread of the electron theory of metals. *Historical Studies in the Physical Sciences* 17: 191–233.
Edge, D. 1983. Is there too much sociology of science? *Isis* 74: 250–56.
Edwords, F. 1980. Why creationism should not be taught as science. *Creation/Evolution* 1: 2–23.
Edwords, F. 1981. Why creationism should not be taught as science. *Creation/Evolution* 3: 6–37.
Edwords, F. 1983. Is it really fair to give creationism equal time? In *Scientists confront creationism*, edited by L. Godfrey, 301–16. New York: Norton.
Eichstaedt, P. 1989a. Furor over fusion. *New Mexican*, 24 May. [In Newsbank Index 53: A10–A11.]
Eichstaedt, P. 1989b. Lab fires queries at fusion prof. *New Mexican*, 19 April. [In Newsbank Index 41: B14.]
Einstein, A. 1990. *The world as I see it: Out of my later years.* New York: Quality Paperback Books.
Eisenstadt, S. 1987. The classical sociology of knowledge and beyond. *Minerva* 25: 77–91.
Eldredge, N. 1982. *The monkey business: A scientist looks at creationism.* New York: Washington Square Press.
Ellis, W. 1986. Creationism in Kentucky: The response of high school biology teachers. In *Science and creation*, edited by R. Hanson, 72–91. New York: Macmillan.
Energy Research Advisory Board. 1989. *Cold fusion research.* Washington, D.C.: Department of Energy.
Etzioni, A., and C. Nunn. 1976. The public appreciation of science in contemporary America. In *Science and its public: The changing relationship*, edited by G. Holton and W. Blanpied, 229–44. Dordrecht: Reidel.
Eve, R., and F. Harrold. 1991. *The creationist movement in modern America.* Boston: Twayne.
Ewing, R., M. Butler, J. Schirber, and D. Ginley. 1989. Negative results and positive artifacts observed in a comprehensive search for neutrons from "cold fusion" using a multidetector system located underground. *Fusion Technology* 16: 404–7.
Fahnestock, J. 1986. Accommodating science: The rhetorical life of scientific facts. *Written Communication* 3, no. 1: 275–96.
Fahnestock, J. 1989. Arguing in different forums: The Bering crossover controversy. *Science, Technology, and Human Values* 14: 26–42.
Fahnestock, J., and M. Secor. 1988. The stases in scientific and literary argument. *Written Communication* 5: 427–43.
Fahnestock, J., and M. Secor. 1991. The rhetoric of literary criticism. In *Textual*

dynamics of the professions, edited by C. Bazerman and J. Paradis, 76–96. Madison: University of Wisconsin Press.

Farrell, L. 1975. Controversy and conflict in science: A case study—the English biometric school and Mendel's laws. *Social Studies of Science* 5: 269–301.

Farrell, T. 1976. Knowledge, consensus, and rhetorical theory. *Quarterly Journal of Speech* 62: 1–14.

Farrell, T. 1978. Social knowledge II. *Quarterly Journal of Speech* 64: 329–34.

Farrell, T., and G. Goodnight. 1981. Accidental rhetoric: The root metaphors of Three Mile Island. *Communication Monographs* 48: 271–300.

Feyerabend, P. 1975a. *Against method: Toward an anarchistic theory of knowledge.* Atlantic Highlands, N.J.: Humanities Press.

Feyerabend, P. 1975b. How to defend society against science. *Radical Philosophy* 11: 3–8.

Feyerabend, P. 1978. From incompetent professionalism to professionalized incompetence: The rise of a new breed of intellectuals. *Philosophy of the Social Sciences* 8: 37–53.

Feyerabend, P. 1979. Dialogue on method. In *The structure and development of science,* edited by G. Radnitzky and G. Andersson, 63–131. Dordrecht: Reidel.

Feyerabend, P. 1985. Imre Lakatos. *British Journal for the Philosophy of Science* 26: 1–18.

Feyerabend, P. 1987. *Farewell to reason.* London: Verso.

Feyerabend, P. 1991. *Three dialogues on knowledge.* London: Basil Blackwell.

Fine, A. 1986. *The shaky game: Einstein, realism, and the quantum theory.* Chicago: University of Chicago Press.

Finocchario, M. 1977. Logic and rhetoric in Lavoisier's sealed note: Toward a rhetoric of science. *Philosophy and Rhetoric* 10: 111–22.

Finocchario, M. 1980. *Galileo and the art of reasoning: Rhetorical foundations of logic and scientific method.* Dordrecht: Reidel.

Fish, S. 1980. *Is there a text in this class?: The authority of interpretive communities.* Cambridge: Harvard University Press.

Fisher, D. 1990. Boundary work and science: The relation between power and knowledge. In *Theories of science in society,* edited by S. Cozzens and T. Gieryn, 98–119. Bloomington: Indiana University Press.

Fiske, J., and J. Hartley. 1987. Bardic television. In *Television: The critical view,* edited by H. Newcombe, 600–612. New York: Oxford University Press.

Fitzpatrick, T. 1989. Pons rejects DOE grant for research. *Salt Lake City Tribune,* 15 April. [In Newsbank Index 40: F6.]

Fitzpatrick, T., and A. Wilson. 1989. U. fusion finding energizes scientists. *Salt Lake City Tribune,* 24 March. [In Newsbank Index 28: E5–E6.]

Fleischmann, M., and S. Pons. 1989. Electrochemically induced nuclear fusion of deuterium. *Journal of Electroanalytical Chemistry and Interfacial Electrochemistry* 261: 301–8.

Fleischmann, M., S. Pons, M. Anderson, L. Li, and M. Hawkins. 1990. Calorimetry of the palladium-deuterium-heavy water system. *Journal of Electroanalytical Chemistry and Interfacial Electrochemistry* 287: 293–348.

Fleischmann, M., S. Pons, and M. Hawkins. 1989. Erratum. *Journal of Electroanalytical Chemistry and Interfacial Electrochemistry* 263: 187–88.

Fleischmann, M., S. Pons, and R. Hoffman. 1989. Measurement of y-rays from cold fusion. *Nature* 339: 667.

Forman, P. 1988. Social niche and self-image of the American physicist. In *The restructuring of the physical sciences in Europe and the United States: 1945–1960,* edited by M. DeMaria, M. Grilli, and F. Sebastiani, 96–104. Singapore: World Scientific Press.

Foucault, M. 1977a. *Discipline and punish: The birth of the prison.* Translated by A. Sheridan. New York: Pantheon.

Foucault, M. 1977b. *Power/knowledge: Selected interviews and other writings.* Translated by C. Gordon. New York: Pantheon.

Foucault, M. 1980. *Power/knowledge.* New York: Pantheon.

Fox, S. 1984. Creationism and evolutionary protobiogenesis. In *Science and creationism,* edited by A. Montagu, 194–239. New York: Oxford University Press.

Franch, J. 1989. Quest for cold fusion still hot debate subject. *Daily Illini,* 8 May, 3.

Franck, R. 1979. Knowledge and opinions. In *Counter movements in the sciences,* edited by H. Nowotny and H. Rose, 39–56. Dordrecht: Reidel.

Frankel, E. 1976. Corpuscular optics and the wave theory of light: The science and politics of a revolution in physics. *Social Studies of Science* 6: 141–84.

Franklin, A., and C. Howson. 1984. Why do scientists prefer to vary their experiments? *Studies in the History and Philosophy of Science* 15: 51–62.

Freeman, D. 1992. A Japanese claim generates new heat. *Science* 256: 438.

French, T. 1989. Getting to the heart of the matter. *St. Petersburg Times,* 23 April. [In Newsbank Index 41: C7–C9.]

Freske, S. 1983. Creationist misunderstanding, misrepresentation, and misuse of the Second Law of Thermodynamics. In *Evolution versus creationism,* edited by J. Zetterberg, 285–95. Phoenix: Oryx Press.

Freudenthal, G. 1979. How strong is Dr. Bloor's "strong programme"? *Studies in the History and Philosophy of Science,* 10: 67–83.

Freudenthal, G. 1984. The role of shared knowledge in science: The failure of the constructivist programme in the sociology of science. *Social Studies of Science* 14: 285–95.

Freudenthal, G. 1987. Joseph Ben-David's sociology of scientific knowledge. *Minerva* 25: 135–49.

Freudenthal, G. 1988. The hermeneutical status of the history of science: The views of Hélène Metzger. In *Science in reflection,* edited by E. Ullman-Margalit, 123–50. Dordrecht: Kluwer.

Fuhrman, E., and K. Oehler. 1986. Discourse analysis and reflexivity. *Social Studies of Science* 16: 293–307.

Fuller, S. 1988. *Social epistemology.* Bloomington: Indiana University Press.

Fuller, S. 1989a. Disciplinary boundaries and the rhetoric of the social sciences. Unpublished MS, Project on the Rhetoric of Inquiry, University of Iowa.

Fuller, S. 1989b. *Philosophy of science and its discontents.* Boulder, Colo.: Westview Press.

Fuller, S. 1992. Social epistemology and the research agenda of science studies. In *Science as practice and culture,* edited by A. Pickering, 390–428. Chicago: University of Chicago Press.

Fuller, S. 1993. *Philosophy, rhetoric, and the end of knowledge: The coming of science and technology studies.* Madison: University of Wisconsin Press.

Fusion energy program: Status and direction. 1989. Hearings before the Subcommittee on Investigations and Oversights of the Committee on Science, Space, and Technology. 101st Congress. Washington, D.C.: GPO.

Futuyma, D. 1983. *Science on trial.* New York: Pantheon.

Gabler, M., and N. Gabler, 1987a. Humanism in textbooks. *Communication Education* 36: 362–66.

Gabler, M., and N. Gabler. 1987b. Moral relativism on the ropes. *Communication Education* 36: 356–61.

Gadamer, H. 1975. *Truth and method.* Translated by G. Barden and J. Cumming. New York: Seabury Press.

Gai, M., S. Rugari, R. France, R. Lund, A. Zhao, A. Davenport, H. Isaacs, and K. Lynn. 1989. Upper limits on neutron and y-ray emission from cold fusion. *Nature* 340: 29–34.

Gallant, R. 1984. To hell with evolution. In *Science and creationism,* edited by A. Montagu, 282–305. New York: Oxford University Press.

Gaonkar, D. 1990. Rhetoric and its double: Reflections on the rhetorical turn in the human sciences. In *The rhetorical turn: Invention and persuasion in the conduct of inquiry,* edited by H. Simons, 341–66. Chicago: University of Chicago Press.

Gaonkar, D. 1993. The idea of rhetoric in the rhetoric of science. *Southern Communication Journal* 58: 258–95.

Gardner, M. 1975. *Fads and fallacies in the name of science.* Toronto: General.

Garfinkel, H., M. Lynch, and E. Livingston. 1981. The work of a discovering science construed with materials from the optically discovered pulsar. *Philosophy of the Social Sciences* 11: 131–58.

Garver, E. 1987. Paradigms and princes. *Philosophy of the Social Sciences* 17: 21–47.

Gaston, J. 1978. The norm of universalism. In *The sociology of science,* edited by J. Gaston, 1–5. San Francisco: Jossey Bass.

Geertz, C. 1968. *Islam observed.* New Haven, Conn.: Yale University Press.

Geertz, C. 1973. *Interpretation of culture.* New York: Basic Books.

Geisler, C. 1991. Toward a sociocognitive model of literacy: Constructing mental models in a philosophical conversation. In *Textual dynamics of the professions,* edited by C. Bazerman and J. Paradis, 171–90. Madison: University of Wisconsin Press.

Geisler, N. 1982. *The Creator in the courtroom: Scopes II.* Milford, Mich.: Mott Media.

Geisler, N., and J. Anderson. 1987. *Origin science: A proposal for the creation-evolution controversy.* Grand Rapids, Mich.: Baker Book House.

Gentry, R. 1984. Radioactive halos in a radiochronological and cosmological perspective. In *Evolutionists confront creationists,* edited by F. Awbrey and

W. Thwaites, 38–65. San Francisco: Pacific Division, American Association for the Advancement of Science.

Gentry, R. 1986. Radioactive halos: Implications for creation. In *Proceedings of the First International Conference on Creationism,* edited by R. Walsh, C. Brooks, and R. Crowell, 89–112. Pittsburgh: Creation Science Fellowship.

Gentry, R. 1987. *Creation's tiny mystery.* Knoxville: Earth Science Associates.

Gergen, D. 1985. The social constructionist movement in modern psychology. *American Psychologist* 40: 266–75.

Gergen, K. 1992. Construction, alienation, and emancipation: Some thoughts on Abir-Am's ethnography of scientific rituals. *Social Epistemology* 6: 365–72.

Giere, R. 1977. History and philosophy of science: Intimate relationship or marriage of convenience? *British Journal for the Philosophy of Science* 28: 282–97.

Giere, R. 1985. Philosophy of science naturalized. *Philosophy of Science* 52: 331–56.

Giere, R. 1988. *Explaining science: A cognitive approach.* Chicago: University of Chicago Press.

Gieryn, T. 1982. Relativist/constructivist programmes in the sociology of science: Redundance and retreat. *Social Studies of Science* 12: 279–97.

Gieryn, T. 1983a. Boundary work and the demarcation of science from non-science: Strains and interests in professional ideologies of scientists. *American Sociological Review* 48: 781–95.

Gieryn, T. 1983b. Making the demarcation of science a sociological problem: Boundary work by John Tyndall, Victorian scientist. In *Working papers on the demarcation of science and pseudoscience,* edited by R. Laudan, 59–86. Blacksburg: Virginia Tech Center for the Study of Science in Society.

Gieryn, T., G. Bevins, and S. Zehr. 1985. Professionalization of American scientists: Public science in the creation/evolution trials. *American Sociological Review* 50: 392–409.

Gieryn, T., and A. Figert. 1986. Scientists protect their cognitive authority: The status degradation of Sir Cyril Burt. In *The knowledge society: The growing impact of scientific knowledge on social relations,* edited by G. Bohme and N. Stehr, 67–86. Dordrecht: Reidel.

Gieryn, T. and A. Figert. 1990. Ingredients for a theory of science in society: O-rings, ice water, Richard Feynman, and the press. In *Theories of science in society,* edited by S. Cozzens and T. Gieryn, 67–97. Bloomington: Indiana University Press.

Gieryn, T., and R. Hirsh. 1983. Marginality and innovation in science. *Social Studies of Science* 13: 87–106.

Gilbert, G., and M. Mulkay. 1980. Contexts of scientific discourse: Social accounting in experimental papers. In *The social process of scientific investigation,* edited by K. Knorr, R. Krohn, and R. Whitley, 269–94. Dordrecht: Reidel.

Gilbert, G., and M. Mulkay. 1981. Putting philosophy to work: Karl Popper's influence on scientific practice. *Philosophy of the Social Sciences* 11: 389–407.

Gilbert, G., and M. Mulkay. 1982. Accounting for error: How scientists construct their social world when they account for correct and incorrect belief. *Sociology* 16: 165–85.

Gilbert, G., and M. Mulkay. 1984. *Opening Pandora's box: A sociological analysis of scientists' discourse.* Cambridge: Cambridge University Press.
Gilkey, L. 1985. *Creationism on trial: Evolution and God at Little Rock.* Minneapolis: Winston.
Gish, D. 1978. *Evolution: The fossils say no!* San Diego: Creation-Life.
Gish, D. 1983. Creation, evolution, and public education. In *Evolution versus creationism,* edited by J. Zetterberg, 175–91. Phoenix: Oryx Press.
Gish, D. 1984. The scientific case for creationism. In *Evolutionists confront creationists,* edited by F. Awbrey and W. Thwaites, 25–37. San Francisco: Pacific Division, American Association for the Advancement of Science.
Gish, D. 1985. *Evolution: The challenge of the fossil record.* El Cajon, Calif.: Master Books.
Gish, D. 1988. Creation, evolution, and the historical evidence. In *But is it science?* edited by M. Ruse, 266–83. Buffalo: Prometheus Books.
Gish, D., R. Bliss, and W. Bird. 1983. Summary of scientific evidence for creation. In *Evolution versus creationism,* edited by J. Zetterberg, 199–207. Phoenix: Oryx Press.
Gittus, J., and J. Bockris. 1989. Explanations of cold fusion. *Nature* 339: 105.
Godfrey, L. 1984. Scientific creationism: The art of distortion. In *Science and creationism,* edited by A. Montagu, 167–81. New York: Oxford University Press.
Godfrey, L. 1985. Footnotes of an anatomist. *Creation/Evolution* 5: 16–36.
Godfrey, L. 1989. Picking a bone with philosophers of science. *Creation/Evolution* 25: 50–53.
Goodnight, G. 1982. The personal, technical, and public spheres of argument: A speculative inquiry into the art of public deliberation. *Journal of the American Forensic Association* 18: 214–27.
Goodnight, G. 1988. The history of rhetoric as a recovery of the public sphere. Paper presented at the annual meeting of the Eastern Communication Association. April.
Goran, M. 1989. *The dangerous ideas of science.* New York: Peter Lang.
Gorman, J. 1981. Creationists vs. evolution. *Discover* 2: 32–34.
Gould, S. 1981. Evolution as fact and theory. *Discover* 2: 33–37.
Gould, S. 1983. Darwin's untimely burial—again! In *Scientists confront creationism,* edited by L. Godfrey, 139–46. New York: Norton.
Gould, S. 1984. Evolution as fact and theory. In *Science and creationism,* edited by A. Montagu, 117–25. New York: Oxford University Press.
Gould, S. 1988. Is a new and general theory of evolution emerging? In *But is it science?* edited by M. Ruse, 177–94. Buffalo: Prometheus Books.
Gould, S., and N. Eldredge. 1977. Punctuated equilibria: The tempo and mode of evolution reconsidered. *Paleobiology* 3: 115–51.
Grabner, I., and W. Reiter. 1979. Guardians at the frontiers of science. In *Counter movements in the sciences,* edited by H. Nowotny and H. Rose, 67–104. Dordrecht: Reidel.
Graham, L. 1978. Concerns about science and attempts to regulate inquiry. *Daedulus* 107: 1–21.

Grave, S. 1960. *The Scottish philosophy of common sense.* Oxford: Oxford University Press.

Green, J. 1985. Media sensationalism and science: The case of the criminal chromosome. In *Expository science: Forms and functions of popularisation,* edited by T. Shinn and R. Whitley, 139–62. Dordrecht: Reidel.

Greenberg, D. 1989. Turf and money—the hot buttons cold fusion pushed. *Chicago Tribune,* 9 May, 17.

Gregson, G., and J. Selzer. 1990. Fictionalizing the readers of scholarly articles in biology. *Written Communication* 7: 25–58.

Gross, A. 1984. Public debates as failed social dramas: The recombinant DNA controversy. *Quarterly Journal of Speech* 70: 397–409.

Gross, A. 1988. On the shoulders of giants: Seventeenth-century optics as an argument field. *Quarterly Journal of Speech* 74: 1–17.

Gross, A. 1989. The rhetorical invention of scientific invention: The emergence and transformation of a social norm. In *Rhetoric in the human sciences,* edited by H. Simons, 89–108. Beverly Hills: Sage.

Gross, A. 1990a. *The rhetoric of science.* Cambridge: Harvard University Press.

Gross, A. 1990b. Rhetoric of science is epistemic rhetoric. *Quarterly Journal of Speech* 76: 304–6.

Gross, A. 1991. Rhetoric of science without constraints. *Rhetorica* 9: 283–99.

Gross, A. 1993. What if we're not producing knowledge? Critical reflections on the rhetorical criticism of science. *Southern Communication Journal* 58: 301–5.

Gross, A. 1994. Guest editor's column. *Technical Communication Quarterly* 3: 5–6.

Gross, A. 1995. Renewing Aristotelian theory: The cold fusion controversy as a test case. *Quarterly Journal of Speech* 81: 48–62.

Grove, J. 1985. Rationality at risk: Science against pseudoscience. *Minerva* 23: 216–40.

Gruenberg, B. 1978. The problem of reflexivity in the sociology of science. *Philosophy of the Social Sciences* 8: 321–43.

Gruenberger, F. 1964. A measure for crackpots. *Science* 145: 1410–14.

Grunbaum, A. 1989. The degeneration of Popper's theory of demarcation. In *Freedom and rationality,* edited by F. D'Agostino and I. Jarvie, 141–61. Dordrecht: Kluwer.

Gusfield, J. 1976. The literary rhetoric of science: Comedy and pathos in drinking driver research. *American Sociological Review* 41: 16–34.

Habermas, J. 1972. *Knowledge and human interests.* London: Heinemann.

Habermas, J. 1973. *Theory and practice.* Boston: Beacon Press.

Hackett, D. 1989. UI researchers work to verify fusion claim. *Champaign-Urbana News Gazette,* 25 April, A2.

Hacking, I. 1979. Imre Lakatos's philosophy of science. *British Journal for the Philosophy of Science* 30: 381–410.

Hacking, I. 1984. *Representing and intervening.* New York: Cambridge University Press.

Hagstrom, W. 1965. *The scientific community.* New York: Basic Books.

Halfpenny, P. 1988. Talking of talking, writing of writing: Some reflections on Gilbert and Mulkay's discourse analysis. *Social Studies of Science* 18: 169–82.

Hall-Jamieson, K. 1987. Protections from censorship. *Communication Education* 36: 402.

Halloran, S. 1978. Technical writing and the rhetoric of science. *Journal of Technical Writing and Communication* 8: 77–88.

Halloran, S. 1984. The birth of molecular biology: An essay in the rhetorical criticism of scientific discourse. *Rhetoric Review* 3: 70–83.

Halstead, L. 1984. Evolution: The fossils say yes! In *Science and creationism,* edited by A. Montagu, 240–54. New York: Oxford University Press.

Hammond, A. and L. Margulis. 1981. Creationism as science. *Science* 81: 55–60.

Handberg, R. 1984. Creationism, conservatism, and ideology: Fringe issues in American politics. *Social Science Journal* 21: 36–51.

Hanson, N. 1958. *Patterns of discovery.* Cambridge: Cambridge University Press.

Hardin, C. 1981. Table turning, parapsychology, and fraud. *Social Studies of Science* 11: 249–55.

Hardin, G. 1984. "Scientific creationism": Marketing deception as truth. In *Science and creationism,* edited by A. Montagu, 159–66. New York: Oxford University Press.

Harding, J. 1982. Establishing scientific guidelines for origins—instruction in public education. In *Creation: The cutting edge,* edited by H. Morris and D. Rohrer, 115–23. San Diego: Creation-Life.

Harding, S. 1986. *The science question in feminism.* Ithaca, N.Y.: Cornell University Press.

Harding, S. 1993. Introduction: Eurocentric scientific illiteracy—a challenge for the world community. In *The racial economy of science: Toward a democratic future,* edited by S. Harding, 1–22. Bloomington: Indiana University Press.

Harding, S., and J. O'Barr, eds. 1987. *Sex and scientific inquiry.* Chicago: University of Chicago Press.

Hargrove, B. 1986. The creationist movement: A sociological view. *Creation/Evolution* 27: 30–38.

Hariman, R. 1986. Status, marginality, and rhetorical theory. *Quarterly Journal of Speech* 72: 38–54.

Hariman, R. 1989. The rhetoric of inquiry and the professional scholar. In *Rhetoric in the human sciences,* edited by H. Simons, 211–32. Beverly Hills: Sage.

Harre, R. 1985. *The philosophies of science: An introductory survey.* Oxford: Oxford University Press.

Harris, R. 1991. Rhetoric of science. *College English* 53: 282–307.

Hays, A. 1957. The Scopes trial. In *Evolution and religion: The conflict between science and theology in modern America,* edited by G. Kennedy, 25–40. Boston: Heath.

Hays, J. 1983. Creation science is not "science": Argument fields and public argument. In *Argument in transition: Proceedings of the Third SCA/AFA Summer Conference on Argumentation,* edited by D. Zarefsky, M. Sillars, and J. Rhodes, 416–22. Annandale, Va.: Speech Communication Association.

Hedtke, R. 1979. An analysis of Darwin's natural selection-artificial selection analogy. *Creation Research Society Quarterly* 16: 89–97, 131.

Heller, A. 1979. Can the unity of sciences be considered as the norm of science? In *Counter movements in the sciences,* edited by H. Nowotny and H. Rose, 57–66. Dordrecht: Reidel.

Hempel, C. 1965. *Aspects of scientific explanation.* New York: Free Press.

Hempel, C. 1966. *Philosophy of natural science.* Englewood Cliffs, N.J.: Prentice-Hall.

Hempel, C., and P. Oppenheim. 1948. Studies in the logic of explanation. *Philosophy of Science* 15: 135–75.

Henry, J. 1986. Occult qualities and the experimental philosophy: Active principles in pre-Newtonian matter theory. *History of Science* 24: 335–81.

Heppenheimer, T. 1984. *The man-made sun: The quest for fusion power.* Boston: Little, Brown.

Herman, R. 1990. *Fusion: The search for endless energy.* Cambridge: Cambridge University Press.

Herrnstein Smith, B. 1994. Circling around, knocking over, playing out. In *Questions of evidence: Proof, practice, and persuasion across the disciplines,* edited by J. Chandler, A. Davidson, and H. Harootunian, 162–68. Chicago: University of Chicago Press.

Hess, D. 1992. The new ethnography and the anthropology of science and technology. *Knowledge and Society* 9: 1–26.

Hesse, M. 1978. Theory and value in the social sciences. In *Action and interpretation: Studies in the philosophy of the social sciences,* edited by C. Hookway and P. Pettit, 4–26. Cambridge: Cambridge University Press.

Hesse, M. 1980. The strong thesis of sociology of science. In *Revolutions and reconstructions in the philosophy of science,* 29–60. Bloomington: Indiana University Press.

Hessen, B. [1931] 1952. The social and economic roots of Newton's *Principia.* In *Science at the cross-roads,* edited by A. Bukharin, 151–212. London: F. Cass.

Hiebert, E. 1986. Modern physics and the Christian faith. In *God and nature: Historical essays on the encounter between science and Christianity,* edited by D. Lindberg and R. Numbers, 424–47. Berkeley: University of California Press.

Hodgson, P. 1979. Presuppositions and limits of science. In *The structure and development of science,* edited by G. Radnitzky and G. Andersson, 133–47. Dordrecht: Reidel.

Hollis, M. 1982. The social destruction of reality. In *Rationality and relativism,* edited by M. Hollis and S. Lukes, 67–86. Cambridge: MIT Press.

Holmquest, A. 1990. The rhetorical strategy of boundary work. *Argumentation* 4: 235–58.

Holton, G. 1978. Sub-electrons, presuppositions, and the Milliken-Ehrenhaft dispute. In *The scientific imagination: Case studies,* 25–83. Cambridge: Cambridge University Press.

Holton, G. 1981. Thematic presuppositions and the direction of scientific advance. In *Scientific explanation,* edited by A. Heath, 1–25. Oxford: Clarendon Press.

Holton, G. 1988. *Thematic origins of scientific thought: Kepler to Einstein.* Rev. ed. Cambridge: Harvard University Press.

Holton, G. 1993. *Science and anti-science.* Cambridge: Harvard University Press.

Holzman, D. 1989. Cold water thrown on fusion. *Washington Times,* 22 May. [In Newsbank Index 52: F8–F9.]

Howard, G. 1985. The role of values in the science of psychology. *American Psychologist* 40: 255–65.

Howe, H., and J. Lyne. 1992a. Gene talk in sociobiology. *Social Epistemology* 6: 109–64.

Howe, H., and J. Lyne. 1992b. Howe and Lyne bully the critics. *Social Epistemology* 6: 231–40.

Huizenga, J. 1992. *Cold fusion: The scientific fiasco of the century.* Rochester, N.Y.: University of Rochester Press.

Hull, D. 1988. *Science as a process: An evolutionary account of the social and conceptual development of science.* Chicago: University of Chicago Press.

Hume, D. 1955. *An inquiry concerning human understanding.* Indianapolis: Bobbs-Merrill.

Hutcheson, P. 1986. Evolution and testability. *Creation/Evolution* 28: 1–8.

Hynes, T. 1989. Can you buy cold fusion by the six pack? or, Bubba and Billy Bob discover Pons and Fleischmann. In *Spheres of argument: Proceedings of the Sixth SCA/AFA Conference on Argumentation,* edited by B. Gronbeck, 42–46. Annandale, Va.: Speech Communication Association.

The ICR scientists. 1982. In *Creation: The cutting edge,* edited by H. Morris and D. Rohrer, 65–75. San Diego: Creation-Life.

Iliffe, R. 1992. Rhetorical vices: Outlines of a Feyerabendian history of science. Review of *Farewell to reason,* by P. Feyerabend. *History of Science* 30: 199–219.

Ilyin, V., and A. Kalinkin. 1985. *The nature of science: An epistemological analysis.* Moscow: Progress.

Iyengar, P., et al. 1990. Bhaba Atomic Research Centre studies in cold fusion. *Fusion Technology* 18: 32–94.

Jackson, P. 1983. A reform of science education: A cautionary tale. *Daedalus* 112: 143–66.

Jacobsen-Wells, J. 1989a. British journal giving U.'s hot new cold shoulder. Salt Lake City *Deseret News,* 3 April. [In Newsbank Index 40: E2–E3.]

Jacobsen-Wells, J. 1989b. Fusion discovery at U. could rank as century's greatest achievement. Salt Lake City *Deseret News,* 24 March. [In Newsbank Index 28: E7–E8.]

Jardine, N. 1989. A dip into the future. *Studies in the History and Philosophy of Science* 20: 15–18.

Jarvie, I. 1983. Rationality and relativism. *British Journal of Sociology* 34: 44–60.

Jarvie, I. 1984. *Rationality and relativism.* Boston: Routledge and Kegan Paul.

Jarvie, I. 1988. Comment. *Current Anthropology* 29: 427–29.

Jasanoff, S. 1987. Contested boundaries in policy-relevant science. *Social Studies of Science* 17: 195–230.

Johnstone, H. 1990. Foreword to *Rhetoric and philosophy,* edited by R. Cherwitz, xv–xviii. Hillsdale, N.J.: Lawrence Erlbaum.

Jones, D. 1987. Realism, rationality, and the rhetoric of science. Paper presented at the annual meeting of the Speech Communication Association, Boston. November.

Jones, R. 1992. *Physics for the rest of us.* Chicago: Contemporary Books.

Jones, S., E. Palmer, J. Czirr, D. Decker, G. Jensen, J. Thorne, S. Taylor, and J. Rafelski. 1989. Observation of cold nuclear fusion in condensed matter. *Nature* 338: 737–40.

Kaufmann, D. 1987. Archaeopteryx and Protoavis. *Creation Research Society Quarterly* 24: 80–82, 185.

Kehoe, A. 1983. The word of God. In *Scientists confront creationism,* edited by L. Godfrey, 1–12. New York: Norton.

Keith, W., and K. Zagacki. 1992. Rhetoric and paradox in scientific revolutions. *Southern Communication Journal* 57: 165–77.

Kelso, J. 1980. Science and the rhetoric of reality. *Central States Speech Journal* 31: 17–29.

Kennedy, G., ed. 1957. *Evolution and religion: The conflict between science and theology in modern America.* Boston: Heath.

Kevles, D. 1977. *The physicists: The history of the scientific community in modern America.* New York: Alfred A. Knopf.

Kirsch, I. 1980. Demonology and science during the scientific revolution. *Journal of the History of the Behavioral Sciences* 26: 359–68.

Kitcher, P. 1982. *Abusing science: The case against creationism.* Cambridge: MIT Press.

Klumpp, J., and T. Hollihan. 1989. Rhetorical criticism as moral action. *Quarterly Journal of Speech* 75: 84–96.

Knorr-Cetina, K. 1981a. *The manufacture of knowledge.* Oxford: Pergamon Press.

Knorr-Cetina, K. 1981b. Social and scientific method; or, What do we make of the distinction between the natural and the social sciences? *Philosophy of the Social Sciences* 11: 335–59.

Knorr-Cetina, K. 1982a. Relativism—what now? *Social Studies of Science* 12: 122–36.

Knorr-Cetina, K. 1982b. Scientific communities or transepistemic arenas of research? A critique of quasi-economic models of science. *Social Studies of Science* 12: 101–30.

Koertge, N. 1972. For and against method. *British Journal for the Philosophy of Science* 23: 274–85.

Kofahl, R., and K. Segraves. 1975. *The creation explanation.* San Diego: Creation-Life.

Kofahl, R., and H. Zeisel. 1981. Popper and Darwinism. *Science* 212: 873.

Kohler, R. 1982. *From medical chemistry to biochemistry: The making of a biomedical discipline.* Cambridge: Cambridge University Press.

Koonin, S., and M. Nauenberg. 1989. Calculated fusion rates in isotopic hydrogen molecules. *Nature* 339: 690–91.

Kottler, M. 1983. Evolution: Fact? theory? . . . or just a theory? In *Evolution versus creationism,* edited by J. Zetterberg, 29–36. Phoenix: Oryx Press.
Kreysa, G., G. Marx, and W. Plieth. 1989. A critical analysis of electrochemical nuclear fusion experiments. *Journal of Electroanalytical Chemistry and Interfacial Electrochemistry* 266: 437–50.
Krige, J. 1978a. A critique of Popper's conception of the relationship between logic, psychology, and a critical epistemology. *Inquiry* 21: 313–35.
Krige, J. 1978b. Popper's epistemology and the autonomy of science. *Social Studies of Science* 8: 287–307.
Krohn, R. 1977. Scientific ideology and scientific process: The natural history of a conceptual shift. In *The social production of scientific knowledge,* edited by E. Mendelsohn, P. Weingart, and R. Whitley, 69–99. Dordrecht: Reidel.
Kuhn, T. 1962. *The structure of scientific revolutions.* Chicago: University of Chicago Press.
Kuhn, T. 1965. Paradigms and some misinterpretations of science. In *Philosophical problems of natural science,* edited by D. Shapere, 83–90. Toronto: Collier-Macmillan.
Kuhn, T. 1977. *The essential tension: Selected studies in scientific tradition and change.* Chicago: University of Chicago Press.
Kuklick, H. 1980. Boundary maintenance in American sociology: Limitations to academic professionalization. *Journal of the History of the Behavioral Sciences* 16: 201–19.
Lachman, S. 1965. *The foundations of science.* New York: Vantage Press.
LaFerriere, J. 1987. Morality, religious symbolism, and the creationist movement. *Creation/Evolution* 21: 1–17.
LaFollette, M. 1982. Science on television: Influences and strategies. *Daedulus* 111: 183–97.
LaFollette, M. 1990. *Making science our own: Public images of science, 1910–1955.* Chicago: University of Chicago Press.
Lakatos, I. 1970. Falsification and the methodology of scientific research programmes. In *Criticism and the growth of knowledge,* edited by I. Lakatos and A. Musgrave, 91–196. Cambridge: Cambridge University Press.
Lakatos, I. 1971. History of science and its rational reconstructions. In *Boston studies in the philosophy of science,* vol. 8, edited by R. Buck and R. Cohen, 91–135. Dordrecht: Reidel.
Lakatos, I. 1978. *The methodology of scientific research programmes: Philosophical papers.* Vol. 1. Cambridge: Cambridge University Press.
Lane, E. 1994. Top scientists hit the unemployment line. *Indianapolis Star,* 31 October, A15.
Lane, J. 1982. Letter. *Physics Today* 35: 15, 103.
Lankford, J. 1981. Amateurs versus professionals: The controversy over telescope size in late Victorian science. *Isis* 72: 11–28.
Larson, E. 1989. *Trial and error: The American controversy over creationism and evolution.* Updated ed. Oxford: Oxford University Press.
Lash, S. 1990. *Sociology of postmodernism.* London: Routledge.

Latour, B. 1983. Give me a laboratory and I will raise the world. In *Science observed: Perspectives on the social study of science,* edited by K. Knorr-Cetina and M. Mulkay, 141–70. London: Sage.

Latour, B. 1987. *Science in action.* Cambridge: Harvard University Press.

Latour, B. 1990. Postmodern? No, simply amodern! Steps toward an anthropology of science. *Studies in the History and Philosophy of Science* 21: 145–71.

Latour, B. 1994. *We have never been modern.* Cambridge: Harvard University Press.

Latour, B., and S. Woolgar. 1979. *Laboratory life: The construction of scientific facts.* Princeton, N.J.: Princeton University Press.

Laudan, L. 1968. Theories of scientific method from Plato to Mach. *History of Science* 7: 1–63.

Laudan, L. 1977. *Progress and its problems.* Berkeley: University of California Press.

Laudan, L. 1981. The pseudo-science of science? *Philosophy of the Social Sciences* 11: 173–98.

Laudan, L. 1982. Two puzzles about science: Reflections on some crises in the philosophy and sociology of science. *Minerva* 20: 253–68.

Laudan, L. 1983. The demise of the demarcation problem. In *Working papers on the demarcation of science and pseudoscience,* edited by R. Laudan, 7–35. Blacksburg: Virginia Tech Center for the Study of Science in Society.

Laudan, L. 1984a. Explaining the success of science: Beyond epistemic realism and relativism. In *Science and reality,* edited by J. Cushing, C. Delaney, and G. Gutting, 83–105. Notre Dame, Ind.: University of Notre Dame Press.

Laudan, L. 1984b. *Science and values.* Berkeley: University of California Press.

Laudan, L. 1986. Some problems facing intuitionist meta-methodologies. *Synthese* 67: 115–29.

Laudan, L. 1987. Relativism, naturalism, and reticulation. *Synthese* 71: 210–34.

Laudan, L. 1988a. More on creationism. In *But is it science?* edited by M. Ruse, 363–66. Buffalo: Prometheus Books.

Laudan, L. 1988b. Science at the bar—causes for concern. In *But is it science?* edited by M. Ruse, 351–55. Buffalo: Prometheus Books.

Laudan, R. 1993. Histories of the sciences and their uses: A review to 1913. *History of Science* 31: 1–34.

Laudan, R., L. Laudan, and A. Donovan. 1988. Testing theories of scientific change. In *Scrutinizing science: Empirical studies of scientific change,* edited by A. Donovan, L. Laudan, and R. Laudan, 3–46. Dordrecht: Kluwer.

Lazarus, B. 1989. Fusion breakthrough could cloud Wyo future. *Casper Star-Tribune,* 16 April. [In Newsbank Index 40: F14.]

Leff, M. 1978. In search of Ariadne's thread: A review of the recent literature on rhetorical theory. *Central States Speech Journal* 29: 73–91.

Leff, M. 1987a. Habitations of rhetoric. In *Argument and critical practices: Proceedings of the fifth SCA/AFA conference on argument,* edited by J. Wenzel, 1–10.

Leff, M. 1987b. Modern sophistic and the unity of rhetoric. In *The rhetoric of the human science: Language and argument in scholarship and public affairs,* edited by J. Nelson, A. Megill, and D. McCloskey, 19–37. Madison: University of Wisconsin Press.

Leggett, A., and G. Baym. 1989. Can solid state effects enhance the cold-fusion rate? *Nature* 340: 45–46.

Lemonick, M. 1989. Fusion illusion? *Time*, 8 May, 72–76.

Lentricchia, F. 1983. *Criticism and social change*. Chicago: University of Chicago Press.

Lepenies, W. 1977. Problems of a historical study of science. In *The social production of scientific knowledge*, edited by E. Mendelsohn, P. Weingart, and R. Whitley, 55–68. Dordrecht: Reidel.

Lepenies, W., and P. Weingart. 1983. Introduction to *Functions and uses of disciplinary histories*, edited by L. Graham, W. Lepenies, and P. Weingart, ix–xx. Dordrecht: Reidel.

Lerner, E. 1991. *The Big Bang never happened*. New York: Random House.

Lerner, E. 1992. The cosmologists' new clothes. *Sky and Telescope* 83: 124.

Lessl, T. 1985. Science and the sacred cosmos: The ideological rhetoric of Carl Sagan. *Quarterly Journal of Speech* 71: 175–87.

Lessl, T. 1988. Heresy, orthodoxy, and the politics of science. *Quarterly Journal of Speech* 74: 18–34.

Lessl, T. 1989. The priestly voice. *Quarterly Journal of Speech* 75: 183–97.

Lewenstein, B. 1992. Cold fusion and hot history. *Osiris* 7: 135–63.

Lewin, R. 1981. A response to creationism evolves. *Science* 217: 635.

Lewin, R. 1985. Evidence for scientific creationism? *Science* 228: 837.

Lewis, R. 1983. Evolution: A system of theories. In *Evolution versus creationism*, edited by J. Zetterberg, 37–64. Phoenix: Oryx Press.

Lewis, S., C. Barnes, M. Heben, A. Kumar, S. Lunt, G. McManis, M. Miskelly, R. Penner, M. Sailor, P. Santangelo, G. Shreve, J. Tufts, M. Youngquist, R. Kavanagh, S. Kellogg, R. Vogelaar, R. Wang, R. Kondrat, and R. New. 1989. Searches for low temperature nuclear fusion of deuterium in palladium. *Nature* 340: 525–30.

Lewontin, R. 1981. Evolution/creationism debate: A time for truth. *Bioscience* 31: 559.

Leys, R. 1981. Meyer's dealings with Jones: A chapter in the history of the American response to psychoanalysis. *Journal of the History of the Behavioral Sciences* 27: 445–65.

Lievrouw, L. 1990. Communication and the social representation of scientific knowledge. *Critical Studies in Mass Communication* 7: 1–10.

Lin, G., J. Kainthla, N. Packham, and M. O'Bockris. 1990. Electrochemical fusion: A mechanism speculation. *Journal of Electroanalytical Chemistry and Interfacial Electrochemistry* 280: 207–11.

Lindley, D. 1989a. More than scepticism. *Nature* 339: 4.

Lindley, D. 1989b. Still no certainty. *Nature* 339: 84.

Lippard, J. 1991–92. How not to argue with creationists. *Creation/Evolution* 29: 9–21.

Losee, J. 1987. *Philosophy of science and historical enquiry*. Oxford: Clarendon Press.

Lucaites, J., and C. Condit. 1993. *Crafting equality.* Chicago: University of Chicago Press.

Lucas, C. 1986. A new unified theory of modern science. In *Proceedings of the First International Conference on Creationism,* edited by R. Walsh, C. Brooks, and R. Crowell, 127–36. Pittsburgh: Creation Science Fellowship.

Lugg, A. 1983. Pseudoscientific practices: Some similarities and differences. In *Working papers on the demarcation of science and pseudoscience,* edited by R. Laudan, 37–57. Blacksburg: Virginia Tech Center for the Study of Science in Society.

Lugg, A. 1987. Bunkum, flim-flam, and quackery: Pseudoscience as a philosophical problem. *Dialectica* 41: 221–30.

Lukes, S. 1973. On the social determination of truth. In *Modes of thought,* edited by P. Horton and M. Finnegan, 230–48. London: Faber.

Lukes, S. 1982. Relativism in its place. In *Rationality and relativism,* edited by M. Hollis and S. Lukes, 261–305. Cambridge: MIT Press.

Lynch, M. 1982. Technical work and critical inquiry: Investigation of a scientific laboratory. *Social Studies of Science* 12: 499–533.

Lynch, M. 1985a. *Art and artifact in laboratory science.* Boston: Routledge and Kegan Paul.

Lynch, M. 1985b. Discipline and the material form of images: An analysis of scientific visibility. *Social Studies of Science* 15: 37–66.

Lynch, M. 1992. Extending Wittgenstein: The pivotal move from epistemology to the sociology of science. In *Science as practice and culture,* edited by A. Pickering, 215–65. Chicago: University of Chicago Press.

Lynch, M., E. Livingston, and H. Garfinkel. 1983. Temporal order in laboratory work. In *Science observed: Perspectives on the social study of science,* edited by K. Knorr-Cetina and M. Mulkay, 205–38. London: Sage.

Lyne, J. 1983. Ways of going public: The projection of expertise in the sociobiology controversy. In *Argument in transition: Proceedings of the Third SCA/AFA Summer Conference on Argumentation,* edited by D. Zarefsky, M. Sillars, and J. Rhodes, 400–415. Annandale, Va.: Speech Communication Association.

Lyne, J. 1985. Rhetorics of inquiry. *Quarterly Journal of Speech* 71: 65–73.

Lyne, J. 1989. Arguing science: The unboundedness of language. In *Spheres of argument: Proceedings of the Sixth SCA/AFA Conference on Argumentation,* edited by B. Gronbeck, 232–38. Annandale, Va.: Speech Communication Association.

Lyne, J. 1990a. Bio-rhetorics: Moralizing the life sciences. In *The rhetorical turn: Invention and persuasion in the conduct of inquiry,* edited by H. Simons, 35–57. Chicago: University of Chicago Press.

Lyne, J. 1990b. The culture of inquiry. *Quarterly Journal of Speech* 76: 192–224.

Lyne, J. 1994. Biorhetorics and public dialogue. Paper presented at the Fourth Public Address Conference, Bloomington, Ind. October.

Lyne, J., and H. Howe. 1986. "Punctuated equilibria": Rhetorical dynamics of a scientific controversy. *Quarterly Journal of Speech* 72: 132–47.

Lyne, J., and H. Howe. 1990. The rhetoric of expertise: E. O. Wilson and sociobiology. *Quarterly Journal of Speech* 76: 134–51.

Lyons, G. 1984. Repealing the Enlightenment. In *Science and creationism*, edited by A. Montagu, 343–62. New York: Oxford University Press.

Lyotard, J. 1984. *The postmodern condition*. Minneapolis: University of Minnesota Press.

MacKenzie, D. 1978. Statistical theory and social interests: A case study. *Social Studies of Science* 8: 35–83.

MacKie, E. 1981. Letter. *Nature* 292: 403.

Maddox, J. 1989a. End of cold fusion in sight. *Nature* 340: 15.

Maddox, J. 1989b. What to say about cold fusion? *Nature* 338: 701.

Maddox, J. 1989c. Where next with peer review? *Nature* 339: 11.

Madison, G. 1990. *The hermeneutics of postmodernity: Figures and themes*. Bloomington: Indiana University Press.

Magyar, G. 1977. Pseudo-effects in experimental physics: Some notes for case studies. *Social Studies of Science* 7: 241–67.

Mainstream scientists respond to creationism. 1982. *Physics Today*, February, 54–59.

Mallove, E. 1991. *Fire from ice: Searching for the truth behind the cold fusion furor*. New York: Wiley.

Mallove, E. 1992. Cold fusion: Real and imaginary. Text of prepared remarks delivered at the University of Utah on 21 May. Engineering and Medical Archives.

Manicas, P., and A. Rosenberg. 1989. The sociology of scientific knowledge: Can we ever get it straight? *Journal for the Theory of Social Behavior* 18: 51–76.

Manier, E. 1989. Reductionist rhetoric: Expository strategies and the development of the molecular neurobiology of behavior. In *The cognitive turn: Sociological and psychological perspectives on science*, edited by S. Fuller, M. De Mey, T. Shinn, and S. Woolgar, 167–98. Dordrecht: Kluwer.

Marcuse, H. 1960. *Reason and revolution*. Boston: Beacon Press.

Marsden, G. 1980. *Fundamentalism and American culture: The shaping of twentieth-century evangelicalism, 1870–1925*. Oxford: Oxford University Press.

Marsden, G. 1984. Understanding fundamentalist views of science. In *Science and creationism*, edited by A. Montagu, 95–116. New York: Oxford University Press.

Martin, B. 1993. The critique of science becomes academic. *Science, Technology, and Human Values* 18: 247–59.

Martin, J. 1989. Ideological critiques and the philosophy of science. *Philosophy of Science* 56: 1–22.

Massey, W. 1989. Science education in the United States: What the scientific community can do. *Science* 245: 915–21.

Maugh, T. 1989a. Cold fusion dispute boils; panelists ridicule claims. *Los Angeles Times*, 3 May. [In Newsbank Index 52: G9–G11.]

Maugh, T. 1989b. Vindication comes to fusion's silent man. *Los Angeles Times*, 30 May. [In Newsbank Index 65: G8–G9.]

Mauskopf, S. 1980. *The reception of unconventional science.* Boulder, Colo.: Westview Press.

May, R. 1984. Creation, evolution, and high school texts. In *Science and creationism* edited by A. Montagu, 306–10. New York: Oxford University Press.

Mazur, A. 1985. *The dynamics of technical controversy.* Washington, D.C.: Communications Press.

McAllister, J. 1986. Theory-assessment in the historiography of science. *British Journal for the Philosophy of Science* 37: 315–33.

McAllister, J. 1992. Competition among scientific disciplines in cold nuclear fusion research. *Science in Context* 5: 17–49.

McCann, H. 1989. Did fusion pair hide essential detail to protect patent claim? *Detroit News,* 12 May. [In Newsbank Index 52: F11.]

McCauley, R. 1986. Intertheoretic relations and the future of psychology. *Philosophy of Science* 53: 179–99.

McCauley, R. 1987. The not so happy story of the marriage of linguistics and psychology; or, Why linguistics has discouraged psychology's recent advances. *Synthese* 72: 341–53.

McCloskey, D. 1985. *The rhetoric of economics.* Madison: University of Wisconsin Press.

McCloskey, D. 1988. The consequences of rhetoric. *Fundamenta Scientiae* 9: 269–84.

McDonald, K. 1991. Cold fusion debate enters a new arena: Publishing houses. *Chronicle of Higher Education,* 20 February, A5, A10.

McDonald, K. 1994. A critical year for nuclear fusion. *Chronicle of Higher Education,* 11 May, A14–A15.

McGee, M. 1980. The ideograph: A link between rhetoric and ideology. *Quarterly Journal of Speech* 66: 1–16.

McGee, M. 1990. Text, context, and the fragmentation of contemporary culture. *Western Journal of Speech Communication* 54: 274–89.

McGee, M., and J. Lyne. 1987. What are nice folks like you doing in a place like this? Some entailments of treating knowledge claims rhetorically. In *The rhetoric of the human sciences: Language and argument in scholarship and public affairs,* edited by J. Nelson, A. Megill, and D. McCloskey, 381–406. Madison: University of Wisconsin Press.

McGhee, L. 1987. The metaphysics of modern science. *Creation Research Society Quarterly* 24: 138–41.

McGuire, J., and T. Melia. 1989. Some cautionary strictures on the writing of the rhetoric of science. *Rhetorica* 7: 87–99.

McGuire, J., and T. Melia. 1991. The rhetoric of the radical rhetoric of science. *Rhetorica* 9: 301–16.

McKerrow, R. 1989. Critical rhetoric: Theory and praxis. *Communication Monographs* 66: 91–111.

McKinlay, A., and J. Potter. 1987. Model discourse: Interpretive repertoires in scientists' conference talk. *Social Studies of Science* 17: 443–63.

McKown, D. 1982. Creationism and the First Amendment. *Creation/Evolution* 7: 24–32.

McLean vs. Arkansas Board of Education. 1982. U.S. District Court, Western Division, Eastern District of Arkansas. NO LR C81, 322.

McMullin, E. 1974. Two faces of science. *Review of Metaphysics* 27: 655–76.

McMullin, E., ed. 1985. *Evolution and creation.* Notre Dame, Ind.: University of Notre Dame Press.

Mehlert, A. 1986. Evolution's revolutions. *Creation Research Society Quarterly* 23: 131–32.

Melia, T. 1984. And lo the footprint . . . Selected literature in rhetoric and science. *Quarterly Journal of Speech* 70: 303–13.

Mendelsohn, E. 1977. The social construction of scientific knowledge. In *The social production of scientific knowledge,* edited by E. Mendelsohn, P. Weingart, and R. Whitley, 3–26. Dordrecht: Reidel.

Merton, R. [1930] 1970. *Science, technology, and society in seventeenth-century England.* New York: Harper and Row.

Merton, R. 1968. *Social theory and social structure.* New York: Free Press.

Merton, R. 1973. *The sociology of science: Theoretical and empirical investigations.* Chicago: University of Chicago Press.

Merton, R. 1976. *Sociological ambivalence.* New York: Free Press.

Meyer, S. 1995. The role of scientists in the "new politics." *Chronicle of Higher Education,* 26 May, B1–B2.

Michael, M. 1992. Lay discourses of science: Science-in-general, science-in-particular, and self. *Science, Technology, and Human Values* 17: 313–33.

Midgley, M. 1985. *Evolution as a religion: Strange hopes and stranger fears.* London: Methuen.

Miles, M., K. Miles, and D. Stilwell. 1990. Electrochemical calorimetric evidence for cold fusion in the palladium-deuterium system. *Journal of Electroanalytical Chemistry and Interfacial Electrochemistry* 296: 241–54.

Miles, M., and R. Miles. 1990. Theoretical neutron flux levels, dose rates, and metal foil activation in electrochemical cold fusion experiments. *Journal of Electroanalytical and Interfacial Electrochemistry* 295: 409–14.

Miller, C. 1989. Some perspectives on rhetoric, science, and history. Review. *Rhetorica* 7: 101–14.

Miller, D. 1986. Method and the "micropolitics" of science: The early years of the geological and astronomical societies of London. In *The politics and rhetoric of scientific method,* edited by J. Schuster and R. Yeo, 227–58. Dordrecht: Reidel.

Miller, J. 1983. Scientific literacy: A conceptual and empirical review. *Daedalus* 112: 29–48.

Miller, J. 1987. Scientific literacy in the United States. In *Communicating science to the public,* edited by D. Evered and M. O'Connor, 19–40. New York: Wiley.

Miller, J., K. Prewitt, and R. Pearson. 1980. *The attitudes of the U.S. public towards science and technology.* Chicago: National Opinion Research Center.

Miller, K. 1983. Answers to the standard creationist arguments. In *Evolution versus creationism,* edited by J. Zetterberg, 249–62. Phoenix: Oryx Press.

Miller, K. 1984. Scientific creationism versus evolution: The mislabeled debate.

In *Science and creationism,* edited by A. Montagu, 18–63. New York: Oxford University Press.
Minkov, J. 1981. Consensual agreement and disagreement in the contemporary scientific community: Some characteristics. *Science of Science* 7: 55–56.
Mitroff, I. 1974. *The subjective side of science: A philosophical inquiry into the psychology of the Apollo moon scientists.* Amsterdam: Elvesier.
Moore, J. 1976. Creationism in California. In *Science and its public: The changing relationship,* edited by G. Holton and W. Blanpied, 191–208. Dordrecht: Reidel.
Moore, J. 1983. Evolution, education, and the nature of science and scientific inquiry. In *Evolution versus creationism,* edited by J. Zetterberg, 3–17. Phoenix: Oryx Press.
Moore, J. 1990–91. Is "creation science" scientific? *Creation/Evolution* 28: 6–15.
Morison, R. 1978. Introduction. *Daedulus* 107: vii–xvi.
Morris, H. 1974. *Many infallible proofs: Practical and useful evidences of christianity.* San Diego: Creation-Life.
Morris, H. 1977. *The scientific case for creation.* San Diego: Creation-Life.
Morris, H. 1980. *The troubled waters of evolution.* San Diego: Creation-Life.
Morris, H. 1982a. An answer for Asimov. In *Creation: The cutting edge,* edited by H. Morris and D. Rohrer, 149–54. San Diego: Creation-Life.
Morris, H. 1982b. The tenets of creationism. In *Creation: The cutting edge,* edited by H. Morris and D. Rohrer, 59–65. San Diego: Creation-Life.
Morris, H. 1984. *A history of modern creationism.* San Diego: Master Book Publishers.
Morris, H. 1989. *The long war against God.* Grand Rapids, Mich.: Baker Book House.
Morris, H. 1991. President's column. *Acts and Facts* 20: 3.
Morris, H., and G. Parker. 1982. *What is creation science?* San Diego: Creation-Life.
Morris, J. 1992. Can scientists study the past? *Acts and Facts* 22: d.
Moss, J. 1989. The interplay of science and rhetoric in seventeenth-century Italy. *Rhetorica* 7: 23–43.
Moyer, W. 1980. Letter. *Bioscience* 30: 4.
Mukerji, C. 1989. *A fragile power: Scientists and the state.* Princeton, N.J.: Princeton University Press.
Mulkay, M. 1975. Norms and ideology in science. *Social Science Information* 15: 535–52.
Mulkay, M. 1979. *Science and the sociology of knowledge.* London: Allen and Unwin.
Mulkay, M. 1981. Action and belief or scientific discourse: A possible way of ending intellectual vassalage in social studies of science. *Philosophy of the Social Sciences* 11: 163–71.
Mulkay, M. 1991. *Sociology of science: A sociological pilgrimage.* Milton Keynes, Eng.: Open University Press.
Mulkay, M., and G. Gilbert. 1982a. Joking apart: Some recommendations concerning the analysis of scientific culture. *Social Studies of Science* 12: 585–614.
Mulkay, M., and G. Gilbert. 1982b. What is the ultimate question? Some remarks in defence of the analysis of scientific discourse. *Social Studies of Science* 12: 309–19.

Mulkay, M., J. Potter, and S. Yearley. 1983. Why an analysis of scientific discourse is needed. In *Science observed: Perspectives on the social study of science,* edited by K. Knorr-Cetina and M. Mulkay, 171–204. London: Sage.
Munevar, G. 1981. *Radical knowledge: A philosophical inquiry into the nature and limits of science.* Indianapolis: Hackett.
Munevar, G. 1985. Rhetorical grounds for determining what is fundamental science: The case of space exploration. In *Argument and social practice: Proceedings of the Fourth SCA/AFA Summer Conference on Argumentation,* edited by J. Cox, M. Sillars, and G. Walker, 420–34. Annandale, Va.: Speech Communication Association.
Murphy, N. 1990. Scientific realism and postmodern philosophy. *British Journal for the Philosophy of Science* 41: 291–303.
Murray, N., and N. Buffaloe. 1983. Creationism and evolution: The real issues. In *Evolution versus creationism,* edited by J. Zetterberg, 454–76. Phoenix: Oryx Press.
Murrell, J. 1993. Hustlers and patrons of science. *History of Science* 31: 65–82.
Musgrave, A. 1968. On a demarcation dispute. In *Problems in the philosophy of science,* edited by I. Lakatos and A. Musgrave, 78–87. Amsterdam: North Holland.
Myers, G. 1985. Texts as knowledge claims: The social construction of two biology articles. *Social Studies of Science* 15: 593–630.
Myers, G. 1989. Persuasion, power, and the conversational model. *Economy and Society* 18: 221–44.
Myers, G. 1990. Sociology of science without the sociology. Review of *A rhetoric of science,* by L. Prelli. *Social Studies of Science* 20: 559–63.
Nelkin, D. 1976. Science or scripture: The politics of "equal time." In *Science and its public: The changing relationship,* edited by G. Holton and W. Blanpied, 209–28. Dordrecht: Reidel.
Nelkin, D. 1977a. Creation vs. evolution: The politics of science education. In *The social production of scientific knowledge,* edited by E. Mendelsohn, P. Weingart, and R. Whitley, 265–88. Dordrecht: Reidel.
Nelkin, D. 1977b. *Science textbook controversies and the politics of equal time.* Cambridge: MIT Press.
Nelkin, D. 1978. Threats and promises: Negotiating the control of research. *Daedulus* 107: 191–209.
Nelkin, D. 1982. *The creation controversy: Science or Scripture in the schools.* Boston: Beacon Press.
Nelkin, D. 1987. *Selling science: How the press covers science and technology.* New York: W. H. Freeman.
Nelson, C. 1986. Creation, evolution, or both? A multiple model approach. In *Science and creation,* edited by R. Hanson, 128–59. New York: Macmillan.
Nelson, J. 1987. Seven rhetorics of inquiry: A provocation. In *The rhetoric of the human sciences: Language and argument in scholarship and public affairs,* edited by J. Nelson, A. Megill, and D. McCloskey, 407–36. Madison: University of Wisconsin Press.
Nelson, J., A. Megill, and D. McCloskey. 1987. Rhetoric of inquiry. In *The rheto-*

ric of the human sciences: Language and argument in scholarship and public affairs, edited by J. Nelson, A. Megill, and D. McCloskey, 3–18. Madison: University of Wisconsin Press.

Newell, N. 1985. *Creation and evolution: Myth or reality?* New York: Praeger.

Newton, I. 1980. Account of the book entituled *Commercium epistolicum.* In *Philosophers at war: The quarrel between Newton and Leibniz,* edited by A. Hall, 263–314. Cambridge: Cambridge University Press.

Newton-Smith, W. 1981. *The rationality of science.* Boston: Routledge and Kegan Paul.

Nicholas, J. 1988. Planck's quantum crisis and shifts in guiding assumptions. In *Scrutinizing science: Empirical studies of scientific change,* edited by A. Donovan, L. Laudan, and R. Laudan, 317–36. Dordrecht: Kluwer.

Nickles, T. 1989. Integrating the science studies disciplines. In *The cognitive turn: Sociological and psychological perspectives on science,* edited by S. Fuller, M. De Mey, T. Shinn, and S. Woolgar, 225–56. Dordrecht: Kluwer.

Nola, R. 1987. The status of Popper's theory of scientific method. *British Journal for the Philosophy of Science* 38: 441–80.

Nola, R. 1990. The strong programme for the sociology of science, reflexivity, and relativism. *Inquiry* 33: 273–96.

Nowotny, H. 1979. Science and its critics: Reflections on anti-science. In *Counter movements in the sciences,* edited by H. Nowotny and H. Rose, 1–26. Dordrecht: Reidel.

Numbers, R. 1982. Creationism in 20th-century America. *Science* 218: 538–44.

Numbers, R. 1985. Science and religion. *Osiris* 1: 59–80.

Numbers, R. 1986. The creationists. In *God and nature: Historical essays on the encounter between science and Christianity,* edited by D. Lindberg and R. Numbers, 391–423. Berkeley: University of California Press.

O'Connor, J., and A. Meadows. 1976. Specialization and professionalization in British geology. *Social Studies of Science* 6: 77–89.

O'Keefe, D. 1975. Logical empiricism and the study of human communication. *Speech Monographs* 42: 169–83.

Olby, R. 1989. The dimensions of scientific controversy: The biometric-Mendelian debate. *British Journal for the History of Science* 22: 299–320.

Olson, R. 1982. *Science deified and science defied: The historical significance in Western culture.* Berkeley: University of California Press.

O'Neill, J. 1981a. The literary production of natural and social science inquiry: Issues and applications in the social organization of science. *Canadian Journal of Sociology* 6: 105–20.

O'Neill, J. 1981b. Marxism and the two sciences. *Philosophy of the Social Sciences* 11: 281–302.

Ono, K., and J. Sloop. 1992. Commitment to telos: A sustained critical rhetoric. *Communication Monographs* 59: 48–60.

Oppenheimer, J. 1966. *Science and the common understanding.* New York: Simon, Schuster.

Oreskes, N. 1988. The rejection of continental drift. *Historical Studies in the Physical Sciences* 18: 311–48.

Ormiston, G., and R. Sassower. 1989. *Narrative experiments: The discursive authority of science and technology.* Minneapolis: University of Minnesota Press.

Orr, C. 1978. How shall we say "Reality is socially constructed through communication"? *Central States Speech Journal* 29: 263–74.

Overington, M. 1977. The scientific community as audience: Toward a rhetorical analysis of science. *Philosophy and Rhetoric* 10: 143–64.

Overn, W., and R. Arndts. 1986. Radiometric dating: An unconvincing art. In *Proceedings of the First International Conference on Creationism,* edited by R. Walsh, C. Brooks, and R. Crowell, 167–74. Pittsburgh: Creation Science Fellowship.

Palfreman, J. 1979. Between scepticism and credulity: A study of Victorian scientific attitudes to modern spiritualism. In *On the margins of science: The social construction of rejected knowledge,* edited by R. Wallis, 201–36. Sociological Review Monograph 27. Keele: University of Keele.

Paolo, P. 1989. Letter. *Nature* 338: 711.

Park, R. 1989. Facts vs. fusion. *Newsday,* 14 May. [In Newsbank Index 52: F2–F6.]

Parker, G. 1980. *Creation: The facts of life.* San Diego: Creation-Life.

Passmore, J. 1978. *Science and its critics.* New Brunswick, N.J.: Rutgers University Press.

Patterson, J. 1983. Thermodynamics and evolution. In *Scientists confront creationism,* edited by L. Godfrey, 99–116. New York: Norton.

Patterson, J. 1984. Thermodynamics and probability. In *Evolutionists confront creationists,* edited by F. Awbrey and W. Thwaites, 132–52. San Francisco: Pacific Division, American Association for the Advancement of Science.

Pearce, D., and V. Rantala. 1984. Scientific change, continuity, and problem solving. In *History of science/philosophy of science: A selection of contributed papers of the Seventh International Congress on Logic, Methodology, and Philosophy of Science,* edited by P. Weingartner and C. Puhringer, 389–99. Meisenheim: Verlag Anton Hain.

Peat, F. 1989. *Cold fusion: The making of a scientific controversy.* Chicago: Contemporary Press.

Perelman, C., and L. Olbrechts-Tyteca. 1969. *The new rhetoric: A treatise on argumentation.* Notre Dame, Ind.: University of Notre Dame Press.

Perelman, D. 1976. Science and the mass media. In *Science and its public: The changing relationship,* edited by G. Holton and W. Blanpied, 246–60. Dordrecht: Reidel.

Perrin, C. 1988. The chemical revolution: Shifts in guiding assumptions. In *Scrutinizing science: Empirical studies of scientific change,* edited by A. Donovan, L. Laudan, and R. Laudan, 105–24. Dordrecht: Kluwer.

Peters, J., and E. Rothenbuler. 1989. The reality of construction. In *Rhetoric in the human sciences,* edited by H. Simons, 11–27. Beverly Hills: Sage.

Peters, T. 1989. On the natural development of public activity: A critique of Goodnight's theory of argument. In *Spheres of argument: Proceedings of the Sixth SCA/AFA Conference on Argumentation,* edited by B. Gronbeck, 26–32. Annandale, Va.: Speech Communication Association.

Petersen, J., ed. 1984. *Citizen participation in science policy.* Amherst: University of Massachusetts Press.

Petrasso, R., X. Chen, K. Wenzel, R. Parker, C. Li, and C. Fiore. 1989a. Petrasso et al. reply. *Nature* 339: 667–69.

Petrasso, R., X. Chen, K. Wenzel, R. Parker, C. Li, and C. Fiore. 1989b. Problems with the y ray spectrum in the Fleischmann et al. experiments. *Nature* 339: 183–85.

Phelps, L. 1982. The dance of discourse: A dynamic, relativistic view of structure. *Pre/Text* 3: 97–142.

Pickering, A. 1980. The role of interests in high energy physics: The choice between color and charm. In *The social process of scientific investigation,* edited by K. Knorr, R. Krohn, and R. Whitley, 107–38. Dordrecht: Reidel.

Pickering, A. 1981. Constraints on controversy: The case of the magnetic monopole. *Social Studies of Science* 11: 69–93.

Pickering, A. 1984a. Against putting the phenomena first: The discovery of the weak neutral current. *Studies in History and Philosophy of Science* 15: 85–117.

Pickering, A. 1984b. *Constructing quarks: A sociological history of particle physics.* Chicago: University of Chicago Press.

Pickering, A. 1988. Big science as a form of life. In *The restructuring of the physical sciences in Europe and the United States: 1945–1960,* edited by M. DeMaria, M. Grilli, and F. Sebastiani, 1–12. Singapore: World Scientific Publishing.

Pickering, A. 1989. Anti-discipline or narratives of illusion. Paper presented at the Dibner Institute Workshop of 20th-Century Science and Technology, Brandeis University, Waltham, Mass. June.

Pickering, A. 1990. Knowledge, practice, and mere construction. *Social Studies of Science* 20: 682–728.

Pickering, A. 1992. From science as knowledge to science as practice. In *Science as practice and culture,* edited by A. Pickering, 1–28. Chicago: University of Chicago Press.

Pickering, A., and A. Stephanides. 1992. Constructing quaternions: On the analysis of conceptual practice. In *Science as practice and culture,* edited by A. Pickering, 139–67. Chicago: University of Chicago Press.

Pinch, T. 1977. What does a proof do if it does not prove? A study of the social conditions and metaphysical divisions leading to David Bohm and John von Neumann failing to communicate in quantum physics. In *The social production of scientific knowledge,* edited by E. Mendelsohn, P. Weingart, and R. Whitley, 171–215. Dordrecht: Reidel.

Pinch, T. 1979. Normal explanations of the paranormal: The demarcation problem and fraud in parapsychology. *Social Studies of Science* 9: 329–48.

Pinch, T. 1985. Towards an analysis of scientific observation: The externality and evidential significance of observational reports in physics. *Social Studies of Science* 15: 3–36.

Pinch, T. 1990. The role of scientific communities in the development of science. *Impact of Science on Society* 40: 219–25.

Pinch, T. 1994. Cold fusion and the sociology of scientific knowledge. *Technical Communication Quarterly* 3: 85–102.

Pinch, T., and H. Collins. 1979. Is anti-science not science? In *Counter movements in the sciences,* edited by H. Nowotny and H. Rose, 221–50. Dordrecht: Reidel.

Pinch, T., and H. Collins. 1984. Private science and public knowledge: The committee for the investigation of the claims of the paranormal and its use of literature. *Social Studies of Science* 14: 521–46.

Pine, R. 1984. But some of them are scientists, aren't they? *Creation/Evolution* 14: 6–18.

Pion, G., and M. Lipsey. 1981. Public attitudes toward science and technology: What have the surveys told us? *Public Opinion Quarterly* 45: 303–16.

Platt, M. 1987. Interpretive strategies/strategic interpretations: On Anglo-American reader-response criticism. In *Postmodernism and politics,* edited by J. Arac, 26–54. Minneapolis: University of Minnesota Press.

Polanyi, M. 1958. *Personal knowledge: Towards a post-critical philosophy.* Chicago: University of Chicago Press.

Pollack, A. 1989. Beating a path to fusion's door. *New York Times,* 28 April, 29, 34.

Pollack, A. 1992. Cold fusion, derided in U.S., is hot in Japan. *New York Times,* 17 November, C1.

Pollner, M. 1991. Left of ethnomethodology: The rise and decline of radical reflexivity. *American Sociological Review* 56: 370–80.

Pollock, R. 1989. In science, error isn't fraud. *New York Times,* 2 May, 27.

Pool, R. 1989a. Cold fusion: End of act I. *Science,* 244: 1039–40.

Pool, R. 1989b. Cold fusion still in state of confusion. *Science* 245: 256.

Pool, R. 1989c. Confirmations heat up cold fusion prospects. *Science* 244: 143–44.

Pool, R. 1989d. Skepticism grows over cold fusion. *Science* 244: 284–85.

Pool, R. 1989e. Teller, Chu boost cold fusion. *Science* 246: 449.

Pool, R. 1989f. Will new evidence support cold fusion? *Science* 246: 206.

Popper, K. 1957. The aim of science. *Ratio* 1: 24–35.

Popper, K. 1959. *The logic of scientific discovery.* London: Hutchinson.

Popper, K. 1962. *Conjectures and refutations.* New York: Harper and Row.

Popper, K. 1968. Remarks on the problems of demarcation and rationality. In *Problems in the philosophy of science,* edited by I. Lakatos and A. Musgrave, 88–101. Amsterdam: North Holland.

Popper, K. 1982. *The open universe: An argument for indeterminism.* Totowa, N.J.: Rowman and Littlefield.

Popper, K. 1988. Darwinism as a metaphysical research program. In *But is it science?* edited by M. Ruse, 144–55. Buffalo: Prometheus Books.

Port, O. 1992. Power in a jar: The debate heats up. *Business Week* 3290: 88–89.

Post, B. 1986. Social problems of science: Trends of interpretation. *Science of Science* 3: 349–60.

Potter, J. 1984. Testability. flexibility: Kuhnian values in scientists' discourse concerning theory choice. *Philosophy of the Social Sciences* 14: 303–30.

Prelli, L. 1989a. The rhetorical construction of scientific ethos. In *Rhetoric in the human sciences,* edited by H. Simons, 48–68. Beverly Hills: Sage.

Prelli, L. 1989b. *A rhetoric of science: Inventing scientific discourse.* Columbia: University of South Carolina Press.

Prelli, L. 1990. Rhetorical logic and the integration of science and rhetoric. *Communication Monographs* 57: 315–22.

Prelli, L. 1993. Rhetorical perspective and the limits of critique. *Southern Communication Journal* 58: 319–27.

Prewitt, K. 1983. Scientific illiteracy and democratic theory. *Daedalus* 112: 49–64.

Price, D. 1961. *Science since Babylon.* New Haven, Conn.: Yale University Press.

Price, D. 1963. *Little science, big science.* New York: Columbia University Press.

Price, D. 1964. The scientific establishment. In *Scientists and national policy making,* edited by R. Gilpin and C. Wright, 19–40. New York: Columbia University Press.

Price, D. 1984. Creationist and fundamentalist apologetics: Two branches of the same tree. *Creation/Evolution* 14: 19–31.

Proctor, D. 1990. The dynamic spectacle: Transforming experience into social forms of community. *Quarterly Journal of Speech* 76: 117–33.

Pumfrey, S. 1991. History of science in the national science curriculum: A critical review of resources and their aims. *British Journal of the History of Science* 24: 61–78.

Putnam, H. 1981. *Reason, truth, and rationality.* Cambridge: Cambridge University Press.

Rabinowitz, M., and D. Worledge 1989. An analysis of cold and lukewarm fusion. *Fusion Technology* 17: 344–49.

Radder, H. 1986. Experiment, technology, and the intrinsic connection between knowledge and power. *Social Studies of Science* 16: 663–83.

Rafelski, J., and S. Jones. 1987. Cold nuclear fusion. *Scientific American* 257: 84–89.

Raths, J. 1973. The emperor's clothes phenomenon in science education. *Journal of Research in Science Teaching* 10: 208–15.

Ratner, S. 1984. Evolution and the rise of the scientific spirit of America. In *Science and creationism,* edited by A. Montagu, 398–415. New York: Oxford University Press.

Raup, D. 1983. The geological and paleontological arguments of creationism. In *Scientists confront creationism,* edited by L. Godfrey, 147–62. New York: Norton.

Raup, D., and S. Stanley. 1978. *Principles of paleontology.* San Francisco: W. H. Freeman.

Recent developments in fusion energy research. 1989. Hearing before the Committee on Space, Science, and Technology. U.S. House of Representatives, 26 April, no. 46.

Reeves, C. 1990. Establishing a phenomenon: The rhetoric of early medical reports on AIDS. *Written Communication* 7: 393–416.

Reichenbach, H. 1938. *Experience and prediction.* Chicago: University of Chicago Press.

Reichenbach, H. 1951. *The rise of scientific philosophy.* Berkeley: University of California Press.

Reid, T. 1966. Essays on the intellectual powers of man. In *18th-century philosophy,* edited by L. White, 134–50. New York: Free Press.

Reid, T. 1970. *An enquiry into the human mind.* Chicago: University of Chicago Press.

Rescher, N. 1984. Extraterrestrial science. In *History of science/philosophy of science: A selection of contributed papers of the Seventh International Congress on Logic, Methodology, and Philosophy of Science,* edited by P. Weingartner and C. Puhringer, 400–424. Meisenheim: Verlag Anton Hain.

Rice, S. 1989. "Faithful in all the little things": Creationists and operation-science. *Creation/Evolution* 25: 8–14.

Richards, R. 1994. Resistance to constructed belief. In *Questions of evidence: Proof, practice, and persuasion across the disciplines,* edited by J. Chandler, A. Davidson, and H. Harootunian, 154–61. Chicago: University of Chicago Press.

Roberts, M. 1976. On the nature and condition of social science. In *Science and its public: The changing relationship,* edited by G. Holton and W. Blanpied, 47–64. Dordrecht: Reidel.

Rolly, P. 1989. Lawyer fees for patents drain funds. *Salt Lake City Tribune,* 3 June. [In Newsbank Index 65: G12–G13.]

Root-Bernstein, R. 1984a. Ignorance versus knowledge in the evolutionist-creationist controversy. In *Evolutionists confront creationists,* edited by F. Awbrey and W. Thwaites, 8–24. San Francisco: Pacific Division, American Association for the Advancement of Science.

Root-Bernstein, R. 1984b. On defining a scientific theory: Creationism considered. In *Science and creationism,* edited by A. Montagu, 64–94. New York: Oxford University Press.

Rorty, R. 1979. *Philosophy and the mirror of nature.* Princeton, N.J.: Princeton University Press.

Rorty, R. 1987. Science and solidarity. In *The rhetoric of the human sciences: Language and argument in scholarship and public affairs,* edited by J. Nelson, A. Megill, and D. McCloskey, 38–52. Madison: University of Wisconsin Press.

Roszak, T. 1976. The monster and the titan: Science, knowledge, and gnosis. In *Science and its public: The changing relationship,* edited by G. Holton and W. Blanpied, 17–32. Dordrecht: Reidel.

Rothman, M. 1990. Cold fusion: A case history in wishful science? *Skeptical Inquirer* 14: 161–70.

Rouse, J. 1987. *Knowledge and power: Toward a political philosophy of science.* Ithaca, N.Y.: Cornell University Press.

Rouse, J. 1991. Philosophy of science and the persistent narratives of modernity. *Studies in the History and Philosophy of Science* 22: 141–62.

Rouse, J. 1993. What are cultural studies of scientific knowledge? *Configurations* 1: 1–22.

Rowland, R. 1986. The relationship between the public and the technical spheres of argument: A case study of the Challenger disaster. *Communication Studies* 37: 136–46.

Rusch, W. 1984. *The argument: Creationism vs. evolutionism.* Terre Haute, Ind.: Creation Research Society Books.

Rusch, W. 1991. *Origins: What is at stake?* Terre Haute, Ind.: Creation Research Society Books.

Ruse, M. 1984. A philosopher's day in court. In *Science and creationism,* edited by A. Montagu, 311–42. New York: Oxford University Press.

Ruse, M. 1988. Pro judice. In *But is it science?* edited by M. Ruse, 356–62. Buffalo: Prometheus Books.

Russman, T. 1987. *A prospectus for the triumph of realism.* Macon, Ga.: Mercer University Press.

Saladin, K. 1986. Educational approaches to creationist politics in Georgia. In *Science and creation,* edited by R. Hanson, 104–27. New York: Macmillan.

Salamon, M., M. Wrenn, H. Bergeson, K. Crawford, D. Delaney, C. Henderson, Y. Li, J. Rusho, G. Sandquist, and S. Seltzer. 1990. Limits on the emission of neutrons, y-rays, electrons, and protons from Pons/Fleischmann electrolytic cells. *Nature* 344: 401–5.

Saltus, R. 1989. On fusion, breakdown instead of breakthrough. *Boston Globe,* 7 May. [In Newsbank Index 52: F14.]

Samelson, F. 1981. Struggle for scientific authority: The reception of Watson's behaviorism, 1913–1920. *Journal of the History of the Behavioral Sciences* 27: 399–425.

Sangren, P. 1988. Rhetoric and the authority of ethnography: Postmodernism and the social reproduction of texts. *Current Anthropology* 29: 405–24.

Sapp, J. 1986. Inside the cell: Genetic methodology and the case of cytoplasm. In *The politics and rhetoric of scientific method,* edited by J. Schuster and R. Yeo, 167–202. Dordrecht: Reidel.

Sarfatti-Larson, M. 1984. The production of expertise and the constitution of expert power. In *The authority of experts,* edited by T. Haskell, 28–83. Bloomington: Indiana University Press.

Sassower, R. 1993. *Knowledge without expertise: On the status of scientists.* Albany: State University of New York Press.

Schadewald, R. 1981–82. Scientific creationism, geocentricity, and the flat earth. *Skeptical Inquirer* 6: 41–48.

Schadewald, R. 1983. The evolution of Bible-science. In *Scientists confront creationism,* edited by L. Godfrey, 301–16. New York: Norton.

Schadewald, R. 1986. Scientific creationism and error. *Creation/Evolution* 6: 1–9.

Schafersman, S. 1983. Fossils, stratigraphy, and evolution: Consideration of a creationist argument. In *Scientists confront creationism,* edited by L. Godfrey, 283–300. New York: Norton.

Schiappa, E. 1989. Spheres of argument as topoi for the critical study of power/knowledge. In *Spheres of argument: Proceedings of the Sixth SCA/AFA Conference on Argumentation,* edited by B. Gronbeck, 47–56. Annandale, Va.: Speech Communication Association.

Schiebinger, L. 1987. The history and philosophy of women and science. In *Sex and scientific inquiry,* edited by S. Harding and J. O'Barr, 7–34. Chicago: University of Chicago Press.

Schmidt, W. 1989. Other laboratories back findings, chemist says. *New York Times,* 18 April, 28.

Schrader-Frechette, K. 1977. Atomism in crisis: An analysis of the current high energy paradigm. *Philosophy of Science* 44: 409–40.

Schrag, C. 1985. Rhetoric resituated at the end of philosophy. *Quarterly Journal of Speech* 71: 164–74.

Schrag, C. 1992. *The resources of rationality: A response to the postmodern challenge.* Bloomington: Indiana University Press.

Schuster, J. 1984. Methodologies as mythic structures: A preface to the historiography of method. *Metascience: Annual Review of the Australasian Association for the History, Philosophy, and Social Studies of Science* 1–2: 15–36.

Schuster, J. 1986. Cartesian method as mythic speech: A diachronic and structural analysis. In *The politics and rhetoric of scientific method,* edited by J. Schuster and R. Yeo, 33–95. Dordrecht: Reidel.

Schwartzman, R. 1989. Nazi racial hygiene and the boundaries of true science. In *Spheres of argument: Proceedings of the Sixth SCA/AFA Conference on Argumentation,* edited by B. Gronbeck, 225–31. Annandale, Va.: Speech Communication Association.

Schwegler, R., and L. Shamoon. 1991. Meaning attribution in ambiguous texts. In *Textual dynamics of the professions,* edited by C. Bazerman and J. Paradis, 216–33. Madison: University of Wisconsin Press.

Scientist is given deadline to prove his research. 1991. *New York Times,* 10 January, C20.

Scott, E., and H. Cole. 1985. The elusive basis of creation science. *Quarterly Review of Biology* 60: 21–30.

Scott, R. 1967. On viewing rhetoric as epistemic. *Central States Speech Journal* 18: 9–16.

Scott, R. 1976. On viewing rhetoric as epistemic: Ten years later. *Central States Speech Journal* 27: 258–66.

Scott, R. 1983. Can a new rhetoric be epistemic? In *The Jensen Lectures: Contemporary communication studies,* edited by J. Sisco, 1–23. Tampa: University of South Florida Department of Communication.

Scully, E. 1984. Review of *Handy dandy evolution refuter,* by R. Kofahl. In *Reviews of thirty-one creationist books,* edited by S. Weinberg, 26–27. Syosset, N.Y.: National Center for Science Education.

Seitz, R. 1989. Fusion in from the cold? *Nature* 339: 185.

Shapere, D. 1965. Introduction to *Philosophical problems of natural science,* edited by D. Shapere, 1–30. Toronto: Collier-Macmillan.

Shapere, D. 1989. Evolution and continuity in scientific change. *Philosophy of Science* 56: 419–37.

Shapin, S. 1979. The politics of observation: Cerebral anatomy and social interests in the Edinburgh phrenology disputes. In *On the margins of science: The social construction of rejected knowledge,* edited by R. Wallis, 139–78. Sociological Review Monograph 27. Keele: University of Keele.

Shapin, S. 1980. Social uses of science. In *The ferment of knowledge: Studies in the historiography of eighteenth-century science,* edited by G. Rosseau and R. Porter, 160–91. Cambridge: Oxford University Press.

Shapin, S. 1982. History of science and its sociological reconstructions. *History of Science* 20: 157–211.

Shapin, S. 1984. Talking history: Reflections on discourse analysis. *Isis* 74: 125–28.

Shapin, S. and S. Schaffer. 1985. *Leviathan and the air-pump.* Princeton, N.J.: Princeton University Press.

Shills, E. 1976. Faith, utility, and the legitimacy of science. In *Science and its public: The changing relationship,* edited by G. Holton and W. Blanpied, 1–16. Dordrecht: Reidel.

Shinn, T. 1986. Failure or success? Interpretations of twentieth-century French physics. *Historical Studies in the Physical Sciences* 16: 353–69.

Shore, S. 1984. Review of *The moon,* by J. C. Whitcomb and D. B. De Young. In *Reviews of thirty-one creationist books,* edited by S. Weinberg, 56–59. Syosset, N. Y.: National Center for Science Education.

Shweber, S. 1986. The empiricist temper regnant: Theoretical physics in the United States, 1920–1950. *Historical Studies in the Physical Sciences* 17: 55–98.

Siitonen, A. 1984. Demarcation of science from the point of view of problems and problem stating. In *History of science/philosophy of science: A selection of contributed papers of the Seventh International Congress of Logic, Methodology, and Philosophy of Science,* edited by P. Weingartner and C. Puhringer, 339–53. Meisenheim: Verlag Anton Hain.

Simons, H. 1980. Are scientists rhetors in disguise? An analysis of discursive processes within scientific communities. In *Rhetoric in transition: Studies in the nature and uses of rhetoric,* edited by E. White, 115–30. University Park: Pennsylvania State University Press.

Simons, H. 1985. Chronicle and critique of a conference. *Quarterly Journal of Speech* 71: 52–64.

Simons, H. 1990. The rhetoric of inquiry as an intellectual movement. In *The rhetorical turn: Invention and persuasion in the conduct of inquiry,* edited by H. Simons, 1–31. Chicago: University of Chicago Press.

Simons, M. 1989. Italians report fusion in experiment. *New York Times,* 19 April, 4.

Sinsheimer, R. 1978. The presumptions of science. *Daedulus* 107: 23–35.

Skoog, G. 1983. The topic of evolution in secondary school biology textbooks: 1900–1977. In *Evolution versus creationism,* edited by J. Zetterberg, 65–89. Phoenix: Oryx Press.

Smart, J. 1972. Science, history, and methodology. *British Journal for the Philosophy of Science* 23: 266–74.

Snow, C. 1981. *The physicists.* Boston: Little, Brown.

Sonleitner, F. 1986. What did Karl Popper really say about evolution? *Creation/Evolution* 28: 9–14.

Sonleitner, F. 1987. The origin of species by punctuated equilibria. *Creation/Evolution* 21: 25–30.

Sorrell, T. 1994. *Scientism: Philosophy and the infatuation with science.* London: Routledge and Kegan Paul.

Staunton, L. 1984. Review of *The origin of the universe,* by Harold Slusher. In *Re-*

views of thirty-one creationist books, edited by S. Weinberg, 52. Syosset, N.Y.: National Center for Science Education.

Stehr, N. 1978. The ethos of science revisited: Social and cognitive norms. In *The sociology of science,* edited by J. Gaston, 172–96. San Francisco: Jossey Bass.

Stehr, N. 1981. The magic triangle: In defense of a general sociology of knowledge. *Philosophy of the Social Sciences* 11: 225–29.

Steinhart, P. 1981. Fundamentals. *Audubon* 83: 142.

Stent, G. 1984. Scientific creationism: Nemesis of sociobiology. In *Science and creationism,* edited by A. Montagu, 136–41. New York: Oxford University Press.

Stewart, J. 1986. Drifting continents and colliding interests: A quantitative application of the interests perspective. *Social Studies of Science* 16: 261–79.

Stokes, T. 1986. Methodology as a normative conceptual problem: The case of the Indian "warped zipper" model of DNA. In *The politics and rhetoric of scientific method,* edited by J. Schuster and R. Yeo, 139–65. Dordrecht: Reidel.

Stowe, D. 1982. *Popper and after: Four modern irrationalists.* Oxford: Pergamon Press.

Strauss, H. 1989a. Ga. Tech learns risks of going public too soon. *Atlanta Journal,* 16 April, [In Newsbank Index 40: F10.]

Strauss, H. 1989b. Scientists skip, stumble along in the chase after cold fusion. *Atlanta Journal,* 25 April, [In Newsbank Index 40: G7.]

Strickling, J. 1979. Creation, evolution, and objectivity. *Creation Research Society Quarterly* 16: 98–101.

Suarez, M., and W. Lemoine. 1986. From internalism to externalism: A study of academic resistance to new scientific findings. *History of Science* 24: 383–410.

Suleiman, S., and I. Crosman, eds. 1980. *The reader in the text: Essays on audience and interpretation.* Princeton, N. J.: Princeton, University Press.

Sullivan, J. 1981. Sociobiology and the crisis of public authority. *Philosophy of the Social Sciences* 11: 271–84.

Suppe, F., ed. 1977. *The structure of scientific theories.* Urbana: University of Illinois Press.

Tarzia, W. 1990. Linguistic tendencies in creationist texts: Hypotheses. *Creation/Evolution* 27: 11–19.

Tate, N. 1989. Fusion claims facing cold reality. *Boston Herald,* 8 May. [In Newsbank Index 52: G8–G9.]

Taubes, G. 1993. *Bad science: The short life and weird times of cold fusion.* New York: Random House.

Taylor, C. 1991a. Defining the scientific community: A rhetorical perspective on communication. *Communication Monographs* 58: 402–20.

Taylor, C. 1991b. Demarcation as argument: The nature and functions of "science" in scientific argument. In *Proceedings of the Second International Conference on Argumentation,* edited by F. Van Eemeren, R. Grootendorst, A. Blair, and C. Willard, 1080–90. Amsterdam: International Society for the Study of Argumentation.

Taylor, C. 1991c. The public closure of technical controversy. In *Argument in controversy: Proceedings of the Seventh SCA/AFA Conference on Argumentation,* edited by D. Parson, 263–69. Annandale, Va.: Speech Communication Association.

Taylor, C. 1992. Of audience, expertise, and authority: The evolving creationism debate. *Quarterly Journal of Speech,* 78: 277–95.
Taylor, C. 1994. Science as cultural practice: A rhetorical perspective. *Technical Communication Quarterly* 3: 67–84.
Taylor, C., and C. Condit. 1988. Objectivity and elites: A creation science trial. *Critical Studies in Mass Communication* 5: 293–312.
Thagard, P. 1980. Resemblance, correlation, and pseudo-science. In *Science, pseudo-science, and society,* edited by M. Hanen, M. Osler, and R. Weyant, 17–23. Waterloo, Ont.: Wilfred Laurier University Press.
Tibbetts, P. 1985. In defense of relativism and the strong programme. *British Journal of Sociology* 36: 471–76.
Tibbetts, P. 1986. "A critical analysis of the sociology of scientific knowledge: Problems of knowledge-legitimation." Ph.D. diss. University of Illinois at Urbana-Champaign.
Tibbetts, P., and P. Johnson. 1985. The discourse and praxis models in recent reconstructions of scientific knowledge generation. *Social Studies of Science* 15: 739–49.
Tolerance but no quarter for creationism. 1981. *Nature* 294: 389–90.
Toulmin, S. 1960. *The philosophy of science.* New York: Harper Torchbooks.
Toulmin, S. 1961. *Foresight and understanding.* New York: Harper Torchbooks.
Toulmin, S. 1965. Ideals of natural order. In *Philosophical problems of natural science,* edited by D. Shapere, 110–23. Toronto: Collier-Macmillan.
Toulmin, S. 1972. *Human understanding: The collective use and evolution of concepts.* Princeton, N.J.: Princeton University Press.
Toulmin, S. 1974. Rationality and scientific discovery. In *Boston studies in the philosophy of science,* vol. 20, edited by K. Shaffner and R. Cohen, 400–21. Dordrecht: Reidel.
Toulmin, S. 1982. The construal of reality: Criticism in modern and postmodern science. *Critical Inquiry* 9: 93–111.
Trachtman, L. 1981. The public understanding of science: A critique. *Science, Technology, and Human Values* 6: 10–15.
Travis, G. 1981. Replicating replication? Aspects of the social construction of learning in planarian worms. *Social Studies of Science* 11: 11–32.
Traweek, S. 1988. *Beamtimes and lifetimes: The world of high energy physics.* Cambridge: Harvard University Press.
Traweek, S. 1992. Big science and colonialist discourse: Building high energy physics in Japan. In *Big science: The growth of large scale research,* edited by P. Galison and B. Hevly, 100–128. Palo Alto, Calif.: Stanford University Press.
Turner, F. 1978. The Victorian conflict between science and religion: A professional dimension. *Isis* 69: 356–76.
Turner, F. 1980. Public science in Britain, 1880–1919. *Isis* 71: 589–608.
Turner, S. 1981. Interpretive charity, Durkheim, and the "strong programme." *Philosophy of the Social Sciences* 11: 231–43.
Turner, S. 1990. Forms of patronage. In *Theories of science in society,* edited by S. Cozzens and T. Gieryn, 185–211. Bloomington: Indiana University Press.
Ullrich, O. 1979. Counter movements and the sciences: Theses supporting

counter movements to the scientisation of the world. In *Counter movements in the sciences*, edited by H. Nowotny and H. Rose, 127–46. Dordrecht: Reidel.
Unger, R. 1989. Basking in glow, or glare, of fusion. *Kansas City Times*, 24 May. [In Newsbank Index 65: F14–G1.]
Urbach, P. 1987. *Francis Bacon's philosophy of science: An account and a reappraisal.* La Salle, Ill.: Open Court Press.
Van, J. 1989a. Naysayers all but put lid on fusion in a jar. *Chicago Tribune*, 10 May, 4.
Van, J. 1989b. Scientists try to cool excitement over fusion. *Chicago Tribune*, 2 May, 4.
Van Den Daele, W. 1977. The social construction of science: Institutionalisation and definition of positive science in the latter half of the seventeenth century. In *The social production of scientific knowledge*, edited by E. Mendelsohn, P. Weingart, and R. Whitley, 27–54. Dordrecht: Reidel.
Van Fraasen, B. 1985. Empiricism in the philosophy of science. In *Images of science*, edited by P. Churchland and C. Hooker, 245–308. Chicago: University of Chicago Press.
Victory for religious freedom! Lawsuit decided in ICR's favor. 1992. *Acts and Facts* 22: 1–3.
Waddell, C. 1989. Reasonableness versus rationality in the construction and justification of science policy decisions: The case of the Cambridge experimentation review board. *Science, Technology, and Human Values* 14: 7–25.
Waddell, C. 1990. The role of pathos in the decision-making process: A study in the rhetoric of science policy. *Quarterly Journal of Speech* 76: 381–400.
Wallace, W. 1989. Aristotelian science and rhetoric in transition: The Middle Ages and Renaissance. *Rhetorica* 7: 7–20.
Walling, C., and J. Simon. 1989. Two innocent chemists look at cold fusion. *Journal of Physical Chemistry* 93: 4693–97.
Wander, P. 1976. The rhetoric of science. *Western Speech Communication* 40: 226–35.
Wartofsky, M. 1980. Introduction to *Science, pseudo-science, and society*, edited by M. Hanen, M. Osler, and R. Weyant, 1–9. Waterloo, Ont.: Wilfred Laurier University Press.
Warwick, A. 1992. Cambridge mathematics and Cavendish physics: Cunningham, Campbell, and Einstein's relativity, 1905–1911. Part I: Uses of theory. *Studies in the History and Philosophy of Science* 23: 625–56.
Watson, G. 1987. Make me reflexive, but not yet: Strategies for managing essential reflexivity in ethnographic discourse. *Journal of Anthropological Research* 43: 29–41.
Weber, M. 1958. *From Max Weber: Essays in sociology.* Translated by H. Gerth and C. Mills. New York: Oxford University Press.
Webster, A. 1979. Scientific controversy and socio-cognitive metonymy: The case of acupuncture. In *On the margins of science: The social construction of rejected knowledge*, edited by R. Wallis, 121–38. Sociological Review Monograph 27. Keele: University of Keele.
Webster, A. 1991. *Science, technology, and society: New directions.* New Brunswick, N.J.: Rutgers University Press.

Weigert, A. 1970. The immoral rhetoric of scientific sociology. *American Sociologist* 5: 111–19.

Weimer, W. 1977. Science as a rhetorical transaction: Toward a nonjustificationist conception of rhetoric. *Philosophy and Rhetoric* 10: 1–29.

Weimer, W. 1979. *Notes on the methodology of scientific research.* Hillsdale, N.J.: Lawrence Erlbaum.

Weimer, W. 1980. For and against method: Reflections on Feyerabend and the foibles of philosophy. *Pre/Text* 1: 161–203.

Weingart, P. 1982. The social assessment of science or the de-institutionalization of the scientific profession. *Science, Technology, and Human Values* 7: 53–55.

Wells, K. 1989. For two scientists, fusion creates fame. *Wall Street Journal,* 27 March, B12.

Wenzel, J. 1963. "The rhetoric of science: A study of comments on technical communication in American scientific societies, 1750–1990." Ph.D. diss., University of Illinois at Urbana-Champaign.

Wenzel, J. 1974. Rhetoric and anti-rhetoric in early American scientific societies. *Quarterly Journal of Speech* 60:328–36.

Westfall, R. 1986a. *The construction of modern science: Mechanisms and mechanics.* Cambridge: Cambridge University Press.

Westfall, R. 1986b. The rise of science and the fall of orthodox Christianity: A study of Kepler, Descartes, and Newton. In *God and nature: Historical essays on the encounter between science and Christianity,* edited by D. Lindberg and R. Numbers, 218–37. Berkeley: University of California Press.

Westman, R. 1986. The Copernicans and the churches. In *God and nature: Historical essays on the encounter between science and Christianity,* edited by D. Lindberg and R. Numbers, 76–113. Berkeley: University of California Press.

Westrum, R. 1977. Social intelligence about anomalies: The case of UFO's. *Social Studies of Science* 7: 271–302.

Westrum, R. 1978. Science and social intelligence about anomalies: The case of meteorites. *Social Studies of Science* 8: 461–93.

Weyant, R. 1980. Protoscience, pseudo-science, metaphors, and animal magnetism. In *Science, pseudo-science, and society,* edited by M. Hanen, M. Osler, and R. Weyant, 77–114. Waterloo, Ont.: Wilfred Laurier University Press.

Whitcomb, J., and H. Morris. 1961. *The Genesis flood.* Nutley, N.J.: Presbyterian and Reformed Publishing Co.

Whitley, R. 1976. Umbrella and polytheistic scientific disciplines and their elites. *Social Studies of Science* 6: 471–97.

Whitley, R. 1977. Changes in the social and intellectual organisation of the sciences: Professionalisation and the arithmetic ideal. In *The social production of scientific knowledge,* edited by E. Mendelsohn, P. Weingart, and R. Whitley, 143–70. Dordrecht: Reidel.

Whitley, R. 1982. The establishment and structure of science as reputational organizations. In *Scientific establishments and hierarchies,* edited by N. Elias, H. Martins, and R. Whitley, 313–58. Dordrecht: Reidel.

Whitley, R. 1985. Knowledge producers and knowledge acquirers: Popularisation as a relation between scientific fields and their publics. In *Expository science: Forms and functions of popularisation,* edited by T. Shinn and R. Whitley, 3–30. Dordrecht: Reidel.

Wiley, E. 1981. Letter to editor. *Nature* 290: 730.

Willard, C. 1983. *Argumentation and the social grounds of knowledge.* Tuscaloosa: University of Alabama Press.

Willard, C. 1985. The science of values and the values of science. In *Argument and social practice: Proceedings of the Fourth SCA/AFA Summer Conference on Argumentation,* edited by J. Cox, M. Sillars, and G. Walker, 435–44. Annandale, Va.: Speech Communication Association.

Willard, C. 1989. *A theory of argumentation.* Tuscaloosa: University of Alabama Press.

Willard, C. 1990. The problem of the public sphere: Three diagnoses. In *Argumentation theory and the rhetoric of assent,* edited by D. Williams and M. Hazen, 135–53. Tuscaloosa: University of Alabama Press.

Williams, D., D. Findley, D. Craston, M. Sense, M. Bailey, S. Croft, B. Hooton, C. Jones, A. Kucernak, J. Mason, and R. Taylor. 1989. Upper bounds on "cold fusion" in electroanalytic cells. *Nature* 342: 375–84.

Williams, E. 1980. Evolution and fluctuations—a creationist evaluation. *Creation Research Society Quarterly* 16: 132–36.

Williams, H. 1989. Two UW students report success with cold fusion. *Seattle Times,* 13 April. [In Newsbank Index 40: G4.]

Williamsburg Charter Foundation. 1988. *The Williamsburg Charter survey on religion and public life.* Washington, D.C.: Williamsburg Charter Foundation.

Wisdom, J. 1987. *Challengeability in modern science.* Aldershot, Eng.: Gower.

Wittgenstein, L. 1961. *Tractatus logico-philosophicus.* London: Routledge and Kegan Paul.

Woodmorappe, J. 1979. An anthology of matters significant to creationism and diluviology: Report I. *Creation Research Society Quarterly* 16: 209–19, 227.

Woodward, A., and D. Elliot. 1987. Evolution and creationism in high school textbooks. *American Biology Teacher* 49: 165–70.

Woolgar, S. 1986. On the alleged distinction between discourse and praxis. *Social Studies of Science* 16: 309–17.

Woolgar, S. 1988a. Comment. *Current Anthropology* 29: 430–31.

Woolgar, S. 1988b. *Science: The very idea.* Chichester, Eng.: Ellis Horwood.

Woolgar, S. 1989a. Representation, cognition, and self: What hope for an integration of psychology and sociology? In *The cognitive turn: Sociological and psychological perspectives on science,* edited by S. Fuller, M. De Mey, T. Shinn, and S. Woolgar, 201–24. Dordrecht: Kluwer.

Woolgar, S. 1989b. What is the analysis of scientific rhetoric for? A comment on the possible convergence between rhetorical and social studies of science. *Science, Technology, and Human Values* 14: 47–49.

Woolgar, S. 1991. The turn to technology in social studies of science. *Science, Technology, and Human Values* 16: 20–50.

Woolgar, S., and D. Pawluch. 1985. Ontological gerrymandering: The anatomy of social problems explanations. *Social Problems* 32: 214–27.

Wormald, B. 1993. *Francis Bacon: History, politics, and science, 1561–1626.* Cambridge: Cambridge University Press.

Wright, P. 1979. A study in the legitimisation of knowledge: The "success" of medicine and the "failure" of astrology. In *On the margins of science: The social construction of rejected knowledge,* edited by R. Wallis, 85–102. Sociological Review Monograph 27. Keele: University of Keele.

Wright, P. 1981. On the boundaries of science in seventeenth-century England. In *Sciences and cultures,* edited by E. Mendelsohn and Y. Elkana, 77–100. Dordrecht: Reidel.

Wynne, B. 1979. Between orthodoxy and oblivion: The normalisation of deviance in science. In *On the margins of science: The social construction of rejected knowledge,* edited by R. Wallis, 67–84. Sociological Review Monograph 27. Keele: University of Keele.

Wynne, B. 1991. Knowledges in context. *Science, Technology, and Human Values* 16: 111–21.

Yandell, K. 1986. Protestant theology and natural science in the twentieth century. In *God and nature: Historical essays on the encounter between science and Christianity,* edited by D. Lindberg and R. Numbers, 448–72. Berkeley: University of California Press.

Yankelovich, D. 1982. Changing public attitudes towards science and the quality of life. *Science, Technology, and Human Values* 39: 23–29.

Yearley, S. 1981. Textual persuasion: The role of social accounting in the construction of scientific arguments. *Philosophy of the Social Sciences* 11: 409–35.

Yearley, S. 1982. The relationship between epistemological and sociological cognitive interests: Some ambiguities underlying the use of interest theory in the study of scientific knowledge. *Studies in History and Philosophy of Science* 13: 353–88.

Yearley, S. 1985. Representing geology: Textual structures in the pedagogical presentation of science. In *Expository science: Forms and functions of popularisation,* edited by T. Shinn and R. Whitley, 79–101. Dordrecht: Reidel.

Yeo, R. 1986. Scientific method and the rhetoric of science in Britain, 1830–1917. In *The politics and rhetoric of scientific method,* edited by J. Schuster and R. Yeo, 259–98. Dordrecht: Reidel.

Yeo, R., and J. Schuster, 1986. Introduction to *The politics and rhetoric of scientific method,* edited by J. Schuster and R. Yeo, ix–xxxviii. Dordrecht: Reidel.

Young, L. 1989. Academia's heavies weigh in, pour cold water on Utah fusion theory. *Baltimore Sun,* 2 May. [In Newsbank Index 52: E10.]

Young, R. 1973. The historiographic and ideological context of the nineteenth-century debate on man's place in nature. In *Changing perspectives in the history of science,* edited by M. Teich and R. Young, 344–438. London: Heinemann.

Zandvoort, H. 1988. Nuclear magnetic resonance and the acceptability of guiding assumptions. In *Scrutinizing science: Empirical studies of scientific change,* edited by A. Donovan, L. Laudan, and R. Laudan, 359–76. Dordrecht: Kluwer.

Zappen, J. 1986. Historical perspectives on the philosophy and the history of science. *Pre/Text* 6: 9–29.

Zeigler, J., T. Zabel, J. Cuomo, V. Brusic, G. Cargill, E. O'Sullivan, and A. Marwick. 1989. Electrochemical experiments in cold fusion. *Physical Review Letters* 62: 2929–32.

Ziman, J. 1968. *Public knowledge: An essay concerning the social dimension of science.* Cambridge: Cambridge University Press.

Ziman, J. 1978. *Reliable knowledge: An exploration of the grounds for belief in science.* Cambridge: Cambridge University Press.

Ziman, J. 1980. *Teaching and learning about science and society.* Cambridge: Cambridge University Press.

Ziman, J. 1981. *Puzzles, problems, and enigmas.* Cambridge: Cambridge University Press.

Ziman, J. 1984. *An introduction to science studies.* Cambridge: Cambridge University Press.

Ziman, J. 1991. Public understanding of science. *Science, Technology, and Human Values* 39: 111–21.

Zimmerman, B. 1993. People's science. In *The racial economy of science: Toward a democratic future,* edited by S. Harding, 440–55. Bloomington: Indiana University Press.

Zimmerman, M. 1987. The evolution-creation controversy: Opinions of Ohio high school biology teachers. *Ohio Journal of Science* 87: 115–25.

Zimmerman, M. 1991–92. A survey of pseudoscientific sentiments of elected officials: A comparison of federal and state legislators. *Creation/Evolution* 29: 26–45.

Zuckerman, H. 1968. Patterns of name ordering among authors of scientific papers: A study of social symbolism and its ambiguity. *American Journal of Sociology* 74: 276–91.

Zuckerman, H. 1970. Stratification in American science. *Sociological Inquiry* 40: 235–57.

Zuckerman, H. 1977. Deviant behavior and social control in science. In *Deviance and social change,* edited by E. Sagarin, 87–138. Beverly Hills: Sage.

Zuckerman, H., and R. Merton. 1971. Patterns of evaluation in science: Institutionalization, structure, and functions of the referee system. *Minerva* 9: 66–100.

INDEX

RHETORIC OF THE HUMAN SCIENCES

Lying Down Together: Law, Metaphor, and Theology
Milner S. Ball

Shaping Written Knowledge: The Genre and Activity of the Experimental Article in Science
Charles Bazerman

Textual Dynamics of the Professions: Historical and Contemporary Studies of Writing in Professional Communities
Charles Bazerman and James Paradis, editors

Politics and Ambiguity
William E. Connolly

The Rhetoric of Reason: Writing and the Attractions of Argument
James Crosswhite

Philosophy, Rhetoric, and the End of Knowledge: The Coming of Science and Technology Studies
Steve Fuller

Machiavelli and the History of Prudence
Eugene Garver

Language and Historical Representation: Getting the Story Crooked
Hans Kellner

The Rhetoric of Economics
Donald N. McCloskey

Therapeutic Discourse and Socratic Dialogue: A Cultural Critique
Tullio Maranhão

The Rhetoric of the Human Sciences: Language and Argument in Scholarship and Public Affairs
John S. Nelson, Allan Megill, and Donald N. McCloskey, editors

What's Left? The Ecole Normale Supérieure and the Right
Diane Rubenstein

Understanding Scientific Prose
Jack Selzer, editor

The Politics of Representation: Writing Practices in Biography, Photography, and Policy Analysis
Michael J. Shapiro

The Legacy of Kenneth Burke
Herbert Simons and Trevor Melia, editors

Defining Science: A Rhetoric of Demarcation
Charles Alan Taylor

The Unspeakable: Discourse, Dialogue, and Rhetoric in the Postmodern World
Stephen A. Tyler

Heracles' Bow: Essays on the Rhetoric and the Poetics of the Law
James Boyd White